The Story of the

Original
Dixieland
Jazz
Band

A Da Capo Press Reprint Series

THE ROOTS OF JAZZ

General Editor : Christopher W. White

The Story of the

Original

Dixieland

Jazz *Band*

H. O. Brunn

DA CAPO PRESS • NEW YORK • 1977

Library of Congress Cataloging in Publication Data

Brunn, Harry O
    The story of the Original Dixieland Jazz Band.

    (The Roots of jazz)
    Reprint of the ed. published by Louisiana State
University Press, Baton Rouge.
    1. Original Dixieland Jazz Band.
[ML3561.J3B8 1977]   785'.06'72   77-3971
ISBN 0-306-70892-2

This Da Capo Press edition of *The Story of the Original
Dixieland Jazz Band* is an unabridged republication of the
first edition published in Baton Rouge, Louisiana in 1960.
It is reprinted by arrangement with Louisiana State University Press.

Published by Da Capo Press, Inc.
A Subsidiary of Plenum Publishing Corporation
227 West 17th Street, New York, N. Y. 10011

The Story of the

Original Dixieland Jazz Band

The Story of the

Original

Jazz

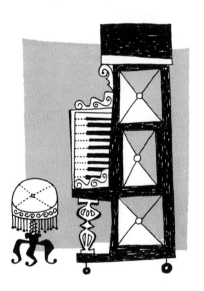

H. O. Brunn

LOUISIANA STATE UNIVERSITY PRESS

Dixieland

Band

Preface

Perhaps no history of an art is more lacking in documentation than that of jazz. The very meaning of the word is obscure to most persons, and even the dictionaries do not agree. The chief complaint to be found with nearly all these definitions is that they are based on hearsay. To complicate matters, many words today carry hardly the same meaning they did yesterday; and those who danced to the jazz that was popular forty years ago would never recognize the music that is presently described by the word.

Many historians speak of jazz as if it had always existed, as if the radical change in popular music around 1916—a change so revolutionary that a new name was required to distinguish it from the ragtime so prevalent throughout the country at that time—never really occurred.

I believe I have presented new and incontrovertible evidence that the Original Dixieland Jazz Band was the first to popularize the radical new music in the leading metropolitan centers of the world, first to increase its spread by means of the world's first jazz phonograph record; and that these men are, despite the attacks levelled at them a few decades after their retirement, more than entitled to the phrase that was always their billing: "The Creators of Jazz."

Information in this book was contributed by many musicians and authorities connected with the Original Dixieland Jazz Band. Foremost of these was the band's leader and organizer, Dominic James (Nick) LaRocca. The interviews with Mr. LaRocca began in New Orleans in the summer of 1946, lasted sometimes for weeks on end, and continued at regular intervals through the present year. Supplementing the personal talks is a correspondence file consisting of more than two hundred of his letters, all relating in detail to the subject of jazz and the Original Dixieland Jazz Band. Most of the photographs reproduced in this book were unearthed from his amazing garage. Here, buried beneath the accumulation of nearly seventy years—everything from 1898 fishing tackle to homemade radio sets of 1927 vintage—were the contracts and records of the historic Original Dixieland Jazz Band. Even Mr. LaRocca himself did not know what to expect, and every box or trunk opened was the source of a new surprise. The documents recovered, having long ago been tossed haphazardly into a trunk or box, were not always in the best state of preservation. Contracts were dog-eared, newspaper clippings yellowed, and photographs torn in half. But the many hundreds of items—account books, ledgers, receipts, passports, letters, fan mail, unpublished compositions—when finally fitted together, formed a rough outline of the band's fascinating past. In 1958 these documents, now known as the LaRocca Collection, were taken over by the Archive of New Orleans Jazz at Tulane University.

These papers established dates, names, and places, but many additional details came from the memories of the musicians themselves. Trombonist Eddie Edwards and drummer Tony Sbarbaro in New York filled in much about the history of the band, and their stories cross-checked to a remarkable degree. Other musicians who at one time or another played with the Original Dixieland Jazz Band—Emile Christian, J. Russel Robinson, Frank Signorelli, Jimmy Lytel, and Henry Levine—contributed valuable details.

One of my greatest regrets was that the great clarinetist of the Dixieland Band, Larry Shields, passed away before making his personal contribution to the book. But Mrs. Shields, now residing in Hollywood, and Harry Shields, a widely respected jazz clarinetist in his own right, were very helpful.

My research in New Orleans was made more pleasant by the genial co-operation of many persons with whom I talked during my periodic visits to that city. The interest in my project shown by such time-honored jazz musicians as Sharkey Bonano, Santo Pecora, and Tom Brown will not soon be forgotten; nor will the efforts of Mrs. Bertha A. Maroney, who came all the way downtown from her job at Ponchartrain Beach to add a few more details to my account of her deceased husband, pianist Henry Ragas of the Original Dixieland Jazz Band.

Special thanks are due Jimmy Durante for the group photograph of his first jazz band and for his occasional notes. I am indebted to Aldo Ricci and Charles R. Iucci of New York Local 802, and to Gus Fischer of Boston Local 9, of the American Federation of Musicians, for their generous help in locating the scores of musicians whose careers had an important bearing on this work. I am also indebted to the many co-operative people at RCA-Victor—among them E. C. Forman, H. C. Darnell, and Louise Sparks—for digging up facts about recordings of the Dixieland Band. Thanks also to Cliff Schweikhard of Buffalo, whose patient and painstaking photography was so essential to the illustration of this book.

Finally, it is important for the reader to be aware of others interested in establishing the Original Dixieland Jazz Band's place in music history, whose efforts along somewhat different lines run parallel to mine.

A West Coast musician and symphony orchestra arranger, Don Fowler, has devoted many years of study to this historic jazz band and has succeeded, where all others have failed, in translating their elusive music onto paper. Working directly from the original phonograph recordings, Fowler created written arrangements which represent the sounds of the band as closely as the limitations of the method will allow. Understandably, the manuscript became discouragingly complex, for the musicians of the Dixieland Band—unable to read music—were not limited to conventional note values and rhythms. Then followed many months of grueling labor in the rehearsal of five musicians competent enough (and durable enough) to follow the intricate scores. The product of their efforts was recorded in Hollywood on an LP disc entitled *Original Dixieland Jazz in Hi-Fi* (ABC-Paramount ABC-184), in which all twelve of the 1917–18 Victor recordings of the Original Dixieland Jazz Band were recreated for modern phonograph reproduction. Although the imitators sometimes fall far short of their masters, the general ensemble effect is often uncanny in its fidelity to the original. The importance of Fowler's project is not so much in its degree of success as in the irrefutable fact that it is the first attempt at dixieland jazz as an *interpretive* rather than a *creative* art. And his record, whatever its shortcomings under close technical scrutiny, comes far closer to describing the music of the Original Dixieland Jazz Band than any other presently available to the public.

In England, my good friend Brian Rust has been instrumental in the reissue of the seventeen sides grooved by the Original Dixieland Jazz Band in that country during 1919 and 1920. The resultant LP records, described and identified elsewhere in this book, were copied from his personal collection and are in some cases clearer than reissues made from original masters. The author of several books and magazine articles on jazz, Rust's research on the Dixieland Band runs closely parallel to my own, with perhaps more emphasis on the recording aspect. He is presently at work on a jazz discography which promises to be by far the most complete in its field, even to the point of listing the matrix numbers of unissued recordings. Likewise, in my own tabulations of phonograph records for this book, I have included unissued as well as issued works. Many of these unreleased masters are still buried in the vaults of disinterested recording companies, and it is optimistically hoped that pressings may some day be made available.

No one who has explored the colorful past of the Original Dixieland Jazz Band can fail to run across the name of J. S. Moynahan. A writer of no small ability, his *Saturday Evening Post* article of February 17, 1937, one of the greatest eulogies of the Original Dixieland Jazz Band ever written, occupies a prominent place in the scrap book of every devoted fan; and his work on the *March of Time* film on the same subject will long be remembered. Moynahan, a jazz clarinetist, and his brother Fred, a drummer, were both active in the early days of the art, and are still among the Dixieland's most loyal disciples.

Last but by no means least, is Henry "Hot Lips" Levine (of "Chamber Music Society of Lower Basin Street" fame), who holds the distinction of having played trumpet in the Original Dixieland Jazz Band during its darkest days, in 1926. His lectures and writings on the subject of dixieland jazz, and his continuance in the medium with his own band in the Miami Beach area, are deserving of special recognition.

To all these good friends and colleagues, this first authentic history of the Original Dixieland Jazz Band is sincerely dedicated.

H. O. Brunn

Snyder, New York
February 16, 1960

Contents

Illustrations

Tables

Prologue

On a bitter cold night in January, 1917, five laughing young men mounted the orchestra stand at Reisenweber's Restaurant in New York City and began an engagement that changed the course of American music. They were the Original Dixieland Jazz Band from New Orleans, and they introduced a music of their own creation—jazz.

As a new form of music, it was revolutionary. It shocked, frightened, confused, and finally captivated the listener. Their first recording of jazz sold over a million copies; the Dixieland Band became nationally famous within a few months. In all parts of the country, the commercial possibilities of their music were immediately recognized by enterprising orchestra leaders, who converted from ragtime to jazz almost overnight. Jazz was exploited at a dizzy pace, and although it existed in its pure form for only a few years, it became the basis for the various styles of dance music that followed.

The phenomenal success of the Dixieland Jazz Band obscured all other bands of the period. The most highly paid dance orchestra in the world, they were noted as an ensemble—their fortune depended upon their original combination. Today, after more than four decades, the Dixieland Band has become almost legendary; and though few people remember anything about it, nearly everyone has heard of it.

Unfortunately for the future of jazz, the Dixieland Band's would-be competitors cared little about its artistic value. The noisier they played, the more popular they became, and vice versa. Imitators were unable to divine the secret that set the original combination apart from all others. New York bandleaders scoured the South and brought north every parade band musician that could blow a flatted seventh. And yet, somehow, the inimitable rhythm of the Original Dixieland Jazz Band could not be duplicated. Every other spurious jazz band for the next two years, as the record shows, was a dismal failure.

The Columbia Gramophone Company, understandably disturbed by the mounting popularity of the Dixieland Band's first Victor records, sent a talent scout to New Orleans in search of a rival jazz band to compete with Victor's great find. But the scout, after combing every café and place of entertainment in the Crescent City, was finally compelled to wire back: "NO JAZZ BANDS IN NEW ORLEANS." Nor was the city ashamed of this lack. In 1918 a *Times-Picayune* editorial disowned the five young upstarts who had started this musical revolution in the North and declined to admit that their fair city had had any part in the making of the raucus new "jass."

In the meantime, the Original Dixieland Jazz Band had carried their innovation to London, where the horrified British refused at first to believe their ears. But by the time of their departure, merely a year later, the five clean-cut lads from Dixie had endeared themselves to all who had heard them. Behind them they left seventeen recordings of jazz, all made in England, and a city full of budding English jazz bands. The British have never forgotten this historic invasion of their homeland by the impassioned little group of American rebels. To this day the arrival of the Dixieland Band in England is celebrated annually by the jazz clubs of that country.

But back home the public psychology that had ensured the world-wide popularity of jazz did not stop with music. The five men that had given the Jazz Age its very name soon found themselves swept away in the mad and furious postwar mode of life. Shimmy, bootleg "hooch," and a new standard of morals had joined hands with jazz music to usher in an era that shocked the older generation. The more civic-minded of them formed vigilance societies and crusades to save their young. Music is always judged by the company it keeps. Jazz, although only an innocent manifestation rather than a cause, was to take full blame. The drive to stamp out this dangerous "immoral" dance music was successful; perhaps few people today would believe that jazz was actually outlawed on Broadway by legal action in 1921.

The time was ripe for another musical revolution, and along came a classical violist with ideas for synthesizing jazz with classical music. The man was Paul Whiteman, and although the era of "symphonic" jazz that he initiated in the middle twenties contributed little to jazz as a creative art, it did much to relieve its reputation. When "hot" music was revived in 1935, at the beginning of the "Swing" Age, most of its severest critics had vanished.

But this was the era of big band swing, and the musician's only freedom as an individual was expressed through his improvised solos, which often became automatic, meaningless exhibitions of technique.

And yet lovers of real jazz in its pure form were not

entirely extinct. They were not great in number, it is true, but with practically no real jazz being played anywhere, no converts or recruits could be enlisted from the new generation. The old school was unorganized, traditionally inarticulate, and not a little overwhelmed and dumbfounded by the inspired exhibitionism of the modern jazz addicts. But there were still enough of the old traditional jazz musicians around to warrant an occasional if meek attempt at pure jazz. This was labeled "dixieland" to distinguish it from the ubiquitous form of music that had usurped the name of "jazz." Tommy Dorsey recognized this faithful minority when he experimented with the Clambake Seven, Bob Crosby found the mark with his Bobcats, and stubborn groups of old-time jazz men were working at the Commodore Music Shop during the war years, turning out a diluted and half-sincere form of dixieland under the Commodore label.

Unfortunately, the Bobcats did not stay "dixie" very long. The original small combination was gradually augmented until finally it equaled the size of its parent orchestra, and the "dixieland band" which Crosby so loudly proclaimed became just another typical swing orchestra playing from written scores. Dorsey gave up the Clambake Seven even earlier. The public was not ready. Dixieland went underground and jazz entered another era.

Then in 1948 an amazing thing happened. Pee Wee Hunt (of all persons!) formed a dixieland combination and put out what he probably considered a tongue-in-cheek version of "Twelfth Street Rag." It was meant to be "corny," to be laughed at, but the coating of humor couldn't disguise the honest jazz underneath. Another generation that had never been educated in what was supposed to be called corny was exposed to a new treat: real, honest jazz. To Pee Wee Hunt's amused amazement, these youngsters went wild. Most of them stopped jitter-bugging long enough to listen. They thought the "ricky-tick" choruses catchy and not at all offensive. Why? Because ricky-tick was brand new to them, and not just a hackneyed and has-been music, as it was to the now-mature proponents of progressive jazz.

Suddenly dixieland began to boom. Every year brought more and more young dixieland combinations to the scene. The Tigertown Five roared out of their den at Princeton, the Salt City Five of Syracuse University began blazing a new (or old) trail across the country, and the Dukes of Dixieland invaded the North with a solid beat. Dixieland clubs were formed, dixieland phonograph records jumped in sales, and even special recording companies were formed to cater to the now-established dixieland clientele.

Surprisingly enough, this "dixieland" was the very same music that was introduced as "jazz" by the Original Dixieland Jazz Band that historic night at Reisenweber's in 1917. Only the audience had changed. Forty years earlier, jazz was appreciated only as a dance music. As such, it aroused dancers as no other music had ever done before. Today, the trends in dancing would render pure jazz almost obsolete, were it not for a new kind of interest shown by the public: they now come to listen. Whether it is played on Bourbon Street in New Orleans or at Nick's in Greenwich Village, the dance floors are today filled with seated audiences. Dixieland jazz has at last come into its own as an American classical music, and it has done so under its own power.

As such, it demands an intelligent criticism and an authentic, detailed account of the five New Orleans men who created it, introduced it, and carried it around the world.

1

No Beer—No Music

Dominic James "Nick" LaRocca was always the central figure of the Dixieland Band, the hub around which this fabulous musical wheel of fortune turned. He was the spirit, the source of power, the driving force of the organization. His inexhaustible supply of energy and his almost unlimited capacity for hard work were as much as anything responsible for the success of the new music he created. This enthusiasm for jazz was transmitted to everyone within hearing range, and especially to the musicians with whom he played.

Music came naturally to young LaRocca, although its course was a hard, rocky, uncharted one at the beginning. His father, Giarolamo LaRocca, had come from Italy in 1876 to settle as a shoemaker at 2022 Magazine Street in a part of New Orleans known as the Irish Channel. Although a talented cornetist himself, he had little respect for professional musicians, whom he called "bums," and was determined to discourage musical amitions in his own children. Above all he wanted them to have the security and independence he himself had sought in the New World. He had chosen a profession for each of them, and when his fourth child, Dominic, was born on April 11, 1889, he decided that this one would be a doctor.

However, Dominic was to be a problem child, for his inclination toward music was apparent before he was old enough to go to school. Using junk he found around the house, he fashioned his own makeshift musical instruments and even drafted his brothers and sisters—Rosario, Antonia, Maria, Bartholomew, and Leonard—into a family orchestra. Nick's first instrument consisted of an empty spaghetti box with bailing wire for strings, and a bow which made use of horsehairs obtained from a nearby livery stable.

Around 1895 Giarolamo LaRocca played cornet at dances and parties; on evenings when he played at the Spanish Fort he always took his family along. They would listen quietly and intently as their father played everything from marches and light opera to mazurkas, polkas, waltzes, schottisches, and cakewalks, but probably the most intent listener of all was little curly-haired Nick. The music he heard was being absorbed by an impressionable young mind and would, through a peculiar process in his memory, someday be expressed in an entirely new form.

On his free evenings Mr. LaRocca would take his children on long walks along the river front. The wharves were not covered or closed in as they are today, and people would stroll along the docks listening to the wide variety of music emanating from the barques or sailing vessels anchored along the way. Every boat seemed to have its own orchestra. There were French, Spanish, Portugese, Italian, and Greek combinations, featuring mostly string instruments but with an occasional accordian, piccolo, or clarinet added. The German bands were mostly brasses, consisting of cornets, alto horns, bass horns, baritone horns, and drums. They often left their boats and passed through the water-front neighborhood, going from corner to corner and passing the hat for contributions.

Sunday afternoon was "open house" on the boats, with soft drinks, wine, cakes, and always music. Nick and his father were often invited to the various vessels, where they would eat, drink, and listen to the music of a dozen nationalities. The more little Nick heard, the more he wanted to be a musician. This was the life of gaiety and happiness, of dancing, laughter, and food aplenty.

It was about this time that Nick began smuggling his father's cornet out of the house and onto the porch of a vacant house on Jackson Avenue, where he made his first attempts at coaxing musical sounds out of the mysterious instrument. Nothing came but air at first; but he had been watching his father, and further attempts at imitating his embouchure finally produced the first encouraging rewards of his experiment—sickening, doleful moans, most likely, but beautiful music to the ears of the young musician. After a few days he had trained his lip to produce bugle calls, and before long he was leading parades down Magazine Street—long processions of kids pounding on wash boilers, old tubs, pie plates, with some carrying broomsticks for guns. Far ahead of the procession, one boy proudly carried an American flag and shouted "Down With Spain!"

Further experiments enabled Nick to find out what the valves were for, and he wrote down the valve numbers as a means of memorizing tunes. When he had rehearsed his first simple melody to his own satisfaction, he approached his father and proudly announced that he could play the cornet. Tingling with pride as he stood before his first audience, the young cornetist blew confidently, his stubby fingers depressing the valves he had so carefully memorized. Father LaRocca listened patiently for the duration of the impromptu concert, then, as little Nick waited for an ovation, he snatched the instrument from the child and smashed it with an axe until it became an unrecognizable mass of flattened brass. From this day forward, the elder LaRocca never again played music. He felt that he had already been an unfavorable influence on his son.

But Nick's compulsion to music was not easily halted. During his summer vacation he earned money by running errands and collecting scrap metal around the neighborhood. At twelve, he had earned enough to buy a used cornet, which he hid at night by tying a string to it and lowering it into the well.

His father rarely saw this instrument, but he heard it often. Nick locked himself in the outhouse and blew the cornet to his heart's delight. He practiced on simple tunes which he played from memory, or learned by writing down the valve numbers on the wall of the outhouse. "There's No Place Like Home" became a part of his early repertoire, but those who lived nearby began to doubt the veracity of the title. Neighborhood kids threw rocks at the outhouse and poured water through the roof, his brothers and sisters turned the hose on the structure and squirted water through the crescent-shaped hole in the door, but it was of no avail—the stubborn cornetist played on.

Within a year he had developed the knack of playing by ear and thereafter dispensed with the habit of writing down valve numbers. The cornet, by this time, had become more than a musical instrument to young Nick. It was more like a close friend, and by far his most valuable and cherished possession. He carried it to St. Alphonsus Parochial School daily, practicing during recess periods and on the way home. But papa La-Rocca, still hoping to save his son from the dismal fate of a professional musician, viewed this situation with mounting alarm.

One day Nick was late for school and hurried off without his precious possession. Returning home that afternoon, he was horrified to find that the second cornet had met the same fate as the first: it had ended up a hopeless mess of scrap metal under the determined axe-blows of his father. The remnant was hung on the door of the outhouse, symbolizing both the degradation and the futility of music as a livelihood.

"I told you before, son," exhorted the elder La-Rocca, "I don't want you to be a musician. Musicians are like gypsies. They play for eats and drink, they wander around the country, they get drunk, they are always penniless. Forget about music, my boy. Study to be a doctor."

But if Giarolamo LaRocca proved anything at all to the neighborhood, it was that horns could be bought faster than they could be destroyed. Further savings, penny by penny, bought a battered old alto horn, which Nick hid in a vacant lot, going there evenings for practice. The valves on this instrument were so corroded they could not be operated, but the relic served to keep his lip in trim.

Mr. LaRocca died shortly after Nick's fifteenth birthday and it became necessary for the boy to leave the University School, a preparatory school he had been attending for three years, for his mother could no longer afford the tuition. He secured a job as extra arc light attendant at the Old French Opera where, in addition to his odd jobs in the daytime, he worked evenings for two seasons, more for the opportunity of watching the singers and musicians than for the pay of a dollar a night. He listened with particular fascination to the operatic fugues. Years later this very musical device was to become an important element in his own compositions.

The first few dollars of Nick's savings went into the down payment on a brand-new silver-plated cornet, with his brother-in-law standing in for the balance, as Nick was still a minor. Although his mother was no more enthusiastic than his father had been, she allowed him to practice in the house. Here he played daily to the accompaniment of a wind-up phonograph and records of Sousa's band. LaRocca's first love was always the military parade band, and the influence of John Philip Sousa is nowhere more apparent than in the structure of his own original compositions, especially the one-steps.

One evening while he was playing along with the player piano, with his two brothers furnishing the manpower, a Mrs. Jacob Young came into the store to purchase shoes for her children. Hearing Nick on the cornet, she told Mrs. LaRocca that her son Henry, who was about Nick's age, was an ardent violinist. She further suggested that Nick spend the summer with Henry at their summer home at Long Beach, Mississippi.

Nick was thrilled at the prospect of playing with

another musician, and it was during the summer of 1905 that he and Henry formed a small orchestra. With Nick on cornet and Henry on violin, the combination also included Joe Guiffre on guitar and an older fellow known simply as "Joe the Barber" on bass viol. This type of orchestra was known as a "no beer—no music" outfit, for although Nick and Henry were both teetotalers, they played for drinks only. They were active almost every night, playing along the Mississippi Gulf Coast in Biloxi, Gulfport, Long Beach, Henderson's Point, Pass Christian, and Log Town.

Henry Young returned to New Orleans with LaRocca the following winter and organized another band in the Magazine Market section with Fonce Price and the Tujague brothers (See Table 1). Between 1906 and 1907 LaRocca played with other string groups, using such local musicians as Sousou Ramos, Willie Guitar, Jim Ruth, and Buzz Harvey.

In most of these ragtime groups the guitar player was the central figure. It was he who would shout out the chord changes on unfamiliar melodies or on modulations to a different key. And the last chorus was known in their trade as the "breakdown," in which each instrumentalist would improvise his own part. It was through this frequent "calling out" of chords by the guitarist that many New Orleans musicians of that day, otherwise totally ignorant of written music, came to recognize their chords by letter and number; and though they could not read music, they always knew the key in which they were playing. This thorough knowledge of chords was one of the most distinguishing features of the New Orleans ragtime musician, who perceived every number as a certain chord progression and was quick to improvise within the pattern.

In 1908 Nick LaRocca formed his first band and expressed for the first time his preference for band instruments, both brass and reed. A valve trombone was played by Jules Cassard, and fifteen-year-old Larry Shields played clarinet. The front line of LaRocca, Cassard, and Shields was supported with a strong rhythm section consisting of Joe Tarranto (guitar), Joe Tujague (bass), and Buddy Rogers (drums).

Is is important to understand that no jazz was being played here or anywhere else in 1908. LaRocca's first band was strictly a ragtime combination, like all the others in New Orleans. The strong prevalence of string instruments up to this point in anything except military bands reflects the spirit and orchestration of ragtime. But, although the rhythm had not yet changed, LaRocca's interest in parade bands had already begun to minimize the string element.

The LaRocca band played mostly for dances and parties and considered themselves lucky to get remuneration in the form of food and liquid refreshments. In this voracious outfit the bass fiddle served double duty as musical instrument and as a sort of epicurean Trojan horse. The musicians cut a porthole in the back of the instrument and furtively stashed away enough sandwiches and cake to last the boys the rest of the week.

The early LaRocca band was handicapped by its amateur status. Its reputation as an outfit that played only for "eats" proved an insurmountable obstacle in the search for paying jobs, and Nick finally turned to other orchestras in order to pick up a few extra dollars.

TABLE 1

Early New Orleans Bands in Which
Nick LaRocca Played Cornet

YEAR	PLACE	PERSONNEL
1905	Gulf Coast Area	Henry Young (violin), Nick La-Rocca (cornet), Joe Guiffre (guitar), "Joe the Barber" (bass)
1906	New Orleans (Magazine Market)	Henry Young, Fonce Price (violins), Nick LaRocca (cornet), Joe Tujague (guitar), Jon Tujague (bass)
1907	New Orleans	Harry Nunez (violin), Nick La-Rocca (cornet), Sousou Ramos (guitar), Willie Guitar (bass)
1907	New Orleans	Fonce Price (violin), Nick LaRocca (cornet), Jim Ruth (guitar), Buzz Harvey (bass)
1907	New Orleans	Fred Englart (violin), Nick La-Rocca (cornet), Joe Tarranto (guitar), John Guiffre (bass)
1908	New Orleans (Magazine Street)	Nick LaRocca (cornet), Larry Shields (clarinet), Jules Cassard (trombone), Joe Tarranto (guitar), Jon Tujagne (bass), Buddy Rogers (drums)
1911	New Orleans (Back of Town)	Dominic Barrocca (guitar), Joe Barrocca (bass), Fred Christian (accordian), Nick LaRocca (cornet), Leon Rapollo (clarinet)
1911	New Orleans (Downtown)	Bill Gallity (valve trombone), Nick LaRocca (cornet), John Pallachais (clarinet), Whit Laucher (guitar), Phil Ray (bass), "Tots" (drums)
1911	New Orleans (Irish Channel)	Henry Brunies (trombone), Merrit Brunies (alto horn), Abbie Brunies

TABLE 1—*Continued*

YEAR	PLACE	PERSONNEL
		(bass), Nick LaRocca (cornet), Joe Tarranto (guitar)
1912	New Orleans	Joe Ellerbusch (trombone), Nick LaRocca (cornet), Ed Roland (clarinet), and others
1912–1916	New Orleans	Jack Laine's Reliance Band: Nick LaRocca (cornet and leader), Alcide Nunez (clarinet), Leonce Mello (trombone), Jack Laine (drums)
1915–1916	New Orleans (Haymarket Café)	Johnny Stein (drums), Nick LaRocca (cornet), Alcide Nunez (clarinet), Leonce Mello (trombone), Henry Ragas (piano)

In 1911 he joined Barrocca's band (See Table 1) which included Leon Rapollo on clarinet. Rapollo was the uncle of the famous clarinetist of the New Orleans Rhythm Kings and a cousin of Nick LaRocca's mother. Like LaRocca's father, Leon Rapollo the elder was an Italian immigrant who had learned music in the Old World tradition. He was strictly a note-reader but could play anything he memorized from a written score.

After a short period with Barrocca's band, followed by a stint with the band of Bill Gallity, LaRocca became acquainted with the populous and remarkable Brunies family. Here he filled the cornet position in a band featuring Henry Brunies (trombone), Merrit Brunies (alto horn), and Abbie Brunies (bass viol). Little Georgie Brunies was just learning to walk about this time, but he was soon to take up the alto horn and later become the famous trombonist of the New Orleans

Rhythm Kings and one of the last of the real tailgate men.[1]

The low pay and the surplus of musicians in New Orleans made music an unprofitable profession. During the daytime LaRocca worked at many trades. He was, in succession, a carpenter, plumber, electrician, and foreman of a demolition company.[2]

It was while LaRocca was working at Tom Gessner's that he met Eddie Edwards. Gessner owned a printing plant at 220 Chartres Street and a stationery store at 609 Canal Street, where his office was located. LaRocca functioned as maintenance man for the plant, starting up the presses in the morning and keeping the equipment running properly during the course of the day. The job left plenty of idle moments, during which he would retreat to the roof of the building with his cornet and shatter the ether with the blaring notes of his original compositions. It is reasonable to assume that such tunes as "Tiger Rag," "Ostrich Walk," and "Livery Stable Blues" were conceived on the hot, tar-covered roof of Gessner's Print Shop, while the presses thrashed out a steady rhythm on the floors below.

One afternoon, when a special problem had arisen

[1] "Tailgate" is the style of trombone counterpoint which has become so charactertistic of dixieland jazz, and which derives its name from the position formerly assumed by the trombonist in the horse-drawn wagon, when brass bands plied the streets of New Orleans. To avoid knocking off the hats of other musicians in the band, the slide trombonist normally sat on the tailgate of the vehicle, pointing his horn aft—hence the term "tailgate trombone."

[2] New Orleans residents may be interested to know that he cleared away such almost-forgotten landmarks as the Maginnis Cotton Warehouse at Lafayette and Magazine streets and the old Library Building at Lafayette and Camp, now occupied by the new Post Office Building.

in connection with his maintenance duties, Nick LaRocca walked over to see his boss at the stationery store. While he was there Tom Gessner, who was Potentate of the local Shrine temple, was talking with a young Mason by the name of Edwards, who had come into the store to discuss plans for the next lodge meeting. Gessner thought it would be a good idea for these two musicians to know each other.

"Nick spends a good part of the day serenading the sea gulls with his horn," said Gessner to Edwards and, with that, a new friendship was born which was to have a telling effect upon the course of musical history.

Edwin B. Edwards was born in New Orleans on May 22, 1891, the son of a salesman. He had begun violin lessons at the age of ten and might have remained an obscure violinist had it not been for a wallet which he found on the sidewalk somewhere along St. Joseph Street in 1906. The wallet contained nearly fifty dollars, a fortune of astronomical proportions to the wide-eyed towhead. The temptation to keep the money was strong, but his conscience was stronger, for Eddie remembered his Sunday School lessons. Returning the money to the owner, whose name and address were given in the wallet, he received a ten-dollar reward, the beginning of a new and wonderful era in his life.

With this ten dollars he ordered a brand-new trombone from the mail-order house of Montgomery Ward, and within a few weeks was assiduously teaching himself the rudiments of his new horn. His method was purely original and relied more on the ear than on the eye. But although he favored the trombone, the violin was more productive of income during his early career. Jobs in theater orchestras provided modest part-time pay, and LaRocca remembers seeing Edwards in the pit of a silent movie theater at the corner of Elysian Fields and Dauphine streets long before the two musicians met.

As his trombone technique improved, the young Edwards found jobs in parades and military band concerts. He played with such outfits as Braun's Naval and Military Band, and Braun's Park Concert Band. At the Tonti Social Club, around 1914, he joined Ernest Giardina's band, a seven-piece ragtime combination consisting of Giardina (violin), Edwards (trombone), Emile Christian (cornet), Achille Bacquet (clarinet), Joe Gerosa (guitar), Eddie Giblin (bass viol), and Tony Sbarbaro (drums). Sbarbaro was eventually to go North to play the drums in the Original Dixieland Jazz Band.

For the most part New Orleans musicians fell into three categories: (1) the "paper men" who could not play by ear, (2) the educated "fakers" who could read or play by ear, as the occasion dictated, and (3) the pure "fakers" who couldn't read a note of music. (See Tables 2 and 3.) As Edwards was primarily a note-reader and LaRocca couldn't read music, it is easy to understand why their paths had never crossed until this memorable day at Gessner's stationery store, even though Eddie lived at the corner of Fourth and Magazine, just four blocks away from Nick's house.

The friendship of LaRocca and Edwards grew rapidly. They formed a partnership, went into the electrical contracting business, and pursued this successfully for nearly two years. Edwards belonged to several business and social organizations and had good contacts, so the partnership never lacked business.

TABLE 2

Ragtime Bands Prominent in New Orleans
Before the Chicago Exodus

FAKERS	READERS AND FAKERS	READERS
Abbie Brunies' band	Frank Christian's band	Charles Christian's band
Joe Lalla's band	Tom Brown's band	Gallagher's band
Bill Gallity's band	The Reliance band	Vince DeCourte's band
Johnny Stein's band	Joe Barrocca's band	Bertucci's band
	John Fischer's band	Bohler's band
	Jack Laine's band	DeDroit's band
		Braun's Military Band
		Emile Tosso's Concert Band

They always took their musical instruments with them. In attics of houses they were wiring, they would quit work periodically to harmonize on cornet and trombone. They did all the wiring themselves, and LaRocca's only complaint was Edwards' inclination to doze off in the warm, quiet attics of these newly constructed homes. Nick would be in the cellar, pounding on the wall and shouting to Eddie to pull up the wires, while Eddie would be fast asleep upstairs. LaRocca recalls with a smile, "That man could sleep anywhere."

Occasionally they would take a day off and go to a vaudeville show, mostly at the insistence of Edwards, who would often sponsor the treat. Here they would pick up ideas in showmanship and discuss plans for a band they would some day organize—a band entirely different from anything in New Orleans. Eddie got Nick into Braun's Military Band as third trumpet. "Play harmony

TABLE 3

*Ragtime Musicians Prominent in New Orleans
Before the Chicago Exodus*

CORNET	TROMBONE	CLARINET
Nick LaRocca	Eddie Edwards	Alcide Nunez
Ray Lopez	Tom "Red" Brown	Gus Mueller
Emile Christian	Jules Cassard	Larry Shields
Frank Christian	Bill Gallity	Achille Bacquet
Manuel Mello	Henry Brunies	John Pallachais
John Lalla	Ricky Toms	Eddie Roland
Joe Lalla	Marcus Kahn	Clem Camp
Richard Brunies	Leonce Mello	Johnny Fischer
Lawrence Vega	Charlie Kirsch	Leon Rapollo
George Barth	George "Happy"	(the elder)
Harry Shannon	Schilling	
Johnny DeDroit	Sigmund Behrenson	
"Doc" Behrenson	Joe Ellerbusch	
Pete Pecoppia		
Freddie Neuroth		
Pete Dietrich		

DRUMS	GUITAR	BASS VIOL
Jack Laine	Joe Tarranto	Jon Tujague
Johnny Stein	Joe Guiffre	Willie Guitar
Tony Sbarbaro	Sousou Ramos	Joe Barrocca
Bill Lambert	Arnold Loyacano	Phil Ray
Diddie Stephens	Joe Tujague	Whit Laucher
"Ragbaby" Stephens	Dominic Barrocca	Steve Brown
Buddy Rogers	Whit Laucher	Bud Loyacano
"Pansy" Laine	Joe Gerosa	Buzz Harvey
	Jim Ruth	George Giblin

Note: Because of the mobility demanded of New Orleans bands, pianists did not become prominent among the musicians who congregated in "Exchange Alley."

but don't play too loud," advised Edwards, "and they'll never know you're a faker." He was right, for the fake was never discovered! Nick knew most of the marches by heart, having played along with the Sousa records.

But the story of LaRocca and Edwards was about to take another fateful turn. The venerable Jack "Papa" Laine was to enter the picture and set the stage for one of the most significant events in the history of jazz.

2

Jack Laine and the "Potato" Men

New Orleans in 1914 was a city bursting with music. Parade bands marched the streets in such large numbers that they sometimes collided. Firemen, policemen, the Army, the Navy, and the Marines all fought for top honors in marching and music, and ward politicians knew that a loud band was the best means of capturing and holding public attention. Evenings found a concert in every park. Old folks danced in the streets to the tunes of street-corner orchestras, and nearly every side-wheeler that plied the Mississippi had its own musical combination.

Music came from a dozen foreign shores on tramp steamers, while almost every day was somebody's holiday—for the Irish St. Patrick's Day; for the Italians the feast of St. Joseph or celebrations of the Italian Society, the Contessa Illena Society, and many others. On July 14 the narrow streets of the French Quarter reverberated with lively orchestras and the dancing of quadrilles as the French celebrated the fall of the Bastille. German bands roamed the city, moving from corner to corner; they made their annual *Volksfest* last a whole week. And of course the colossal Mardi Gras parades were the biggest event of all. There was music at picnics, outings, and conventions; at weddings, birthdays, and funerals; at prize fights, ball games, and track meets; at nickelodeons, cafés, and beer gardens. When a New Orleans musician advertised "music for all occasions," he was prepared for anything.

With so much music around, it is not difficult to understand why so many New Orleans kids grew up to be musicians. Every healthy, red-blooded boy in the city, if he didn't want to be a fireman, certainly wanted to be a musician. Before the days of radio and television, when even the phonograph was considered a screechy toy, if you wanted music you made it yourself; these youngsters of the Crescent City emulated their elders by making their own instruments and practicing for the day when they, too, could take their places in the glorious and exciting New Orleans parades.

The more talented ones realized their childhood ambitions. The pride of being a parade band musician was nowhere more boldly expressed than in the uniforms. Young bands that were hardly able to afford carfare managed to scrape up enough change to buy secondhand policemen's coats and firemen's dress hats, to which braided patches would be attached, bearing the name of the bandleader. The coats would be worn with any matching blue trousers that the musicians happened to own.

The meeting place for musicians was Exchange Place, between Canal and Bienville streets, especially Paul Blum's Cafe at 116–118. Here young musicians of the day would congregate to obtain jobs paying upwards of one dollar.

The best contact for a job at this time was a forty-one-year-old parade band drummer by the name of Jack

"Papa" Laine. Laine managed five different bands at once and there was enough demand to keep them all simultaneously active. The twelve-piece Reliance Band was the most famous; this was a military band hired out primarily for parades and outings. The others were smaller combinations—all predominantly reed, brass, and percussion—and were used for parties, picnics, and ballyhoo purposes.

On special occasions, such as Mardi Gras and other festivals, Laine would form several extra parade bands. Although there were hundreds of New Orleans musicians, on these special days the jobs would outnumber the players. Laine would sometimes organize marching bands of ten or twelve men in which only eight or nine would be actual musicians. The others would be "dummies"—drifters whom he hired merely to march and carry instruments. To keep these non-musicians from blowing on their borrowed instruments and making non-musical noises, potatoes or rags were stuffed into the bells of the horns, hence the term "potato" men.

LaRocca and Edwards often came to "Exchange Alley," as it was then called, to hang around, chat with the other musicians, and pick up an occasional job. They both played in the Reliance Band at one time or another, and LaRocca became a regular member. It was Jack Laine who first recognized the ability of young LaRocca and made him leader of his Number 1 band. At other times he would tell LaRocca to appear at a certain place, then would send other musicians there to meet him. The resulting band would always be an unknown quantity, for Nick would never know who was going to show up—if anybody—or what their capabilities would be.

The most memorable example was New Year's Eve of 1915. Laine had sent LaRocca to the Woodmen of the World Hall, Almonaster and Urquhart streets, promising that six other musicians would meet him there for a job at a nearby party. LaRocca waited on the corner for two hours and the only musician who showed up was a bass player by the name of Willie Guitar. Finding available musicians anywhere on New Year's Eve was like looking for snowhoes in the Sahara. So the two musicians took the job and played for six straight hours— just a cornet with bass viol accompaniment!

On December 13, 1915, one of Laine's ballyhoo bands was playing on the corner of Canal and Royal streets, advertising a coming prize fight between Eddie Coulon and the popular New Orleans featherweight, Pete Herman, and receiving a total of $7.25 for their services. Standing in the crowd was a dapper young man named Harry James, a Chicago café owner who had come to New Orleans to witness the very same prize fight now being loudly heralded by Laine's band.[1] James, who had been passing down Canal Street, had been attracted first by this unusually loud and unconventionally fast combination. Now he stood enraptured, absent-mindedly fingering his diamond cuff links as he

[1] Pete Herman, born in New Orleans in 1896 and managed by Sammy Goldman, was one of the most promising young boxers in the city at that time. He was a special favorite of James, who had come all the way from Chicago to see this match. As the contest with Eddie Coulon was Herman's only local fight of the season, it definitely establishes December 13 as the date on which Laine's band was first heard by Harry James.

listened to a kind of music he had never heard before. LaRocca, pointing his cornet skyward and blowing to the point of apoplexy, ripped off the polyphonic phrases of his own original melodies, as trombonist Leonce Mello, seated on the tailgate of the horse-drawn wagon, answered in powerful blatting tones that rattled the plate glass windows across the street. Over this pair rode the screaming clarinet of Alcide Nunez, and behind it the booming of Jack Laine's big parade drum.

James listened for nearly an hour, feeling the driving, relentless beat of this obscure little band, carefully appraising the reactions of the passers-by. Between numbers, while the young musicians were laughing and shoving one another about on top of the wagon, he spoke to Jack Laine, who had momentarily stepped down to the street. He asked him if he would be interested in taking his band to Chicago. Laine explained that his local commitments prevented him from leaving town. But he called LaRocca down from the wagon and told him the story. LaRocca said to James, "If you want to hear some *real* music, come on over to the Haymarket Café tonight after the fight."

That night Harry James watched the fight at the New Orleans Coliseum impassively, for he was still hearing the band on the wagon at Canal and Royal. Herman scored a thrilling four-round knockout over Coulon, but James hardly noticed. The spirited music that had been ringing in his ears all afternoon continued to haunt him. As he fought his way through the boisterous crowd after the main bout, he thought he heard the blaring brass of LaRocca's cornet. He paused, cupped a hand to one ear, but all he heard was the constant babble of human voices. Then, a few minutes later, he heard it again—a single, isolated, syncopated phrase—and followed in its direction until the sounds became more frequent and finally led him into the ramshackle Haymarket Café.

Here was the whole band—LaRocca on cornet, Mello on trombone, Nunez on clarinet, Johnny Stein on drums, and Henry Ragas on piano. The music was ragtime, and there was a lot of it in New Orleans, but there was something different about this bunch. Harry James remained at the café until Stein's band played its last number at half past three.

It seems hardly questionable that the cornet of Nick LaRocca was the outstanding feature of the two bands heard by James that eventful day in 1915. In Laine's band at Canal and Royal, LaRocca was the only future member of the Original Dixieland Jazz Band then present. In Stein's band at the Haymarket Café, pianist Henry Ragas was the only addition. As Nunez and Mello both played melody most of the time, the conventional three-voiced pattern of the typical dixieland combination certainly could not have been present. Therefore it could not have been a distinctive ensemble that the gentleman from Chicago found stimulating.

Harry James returned to Chicago. But he never forgot the fast, raucous, blaring little band that had set his toes tapping uncontrollably on that memorable one-day visit to the Crescent City. He listened to the soupy string ensembles that were bleeding the life out of ragtime in Chicago, and he tried to visualize these five southern lads on his bandstand at the Booster's Club in the Hotel Morrison. It seemed mad at first thought. A brass band

at a honky-tonk café or outdoor ballyhoo in New Orleans
was one thing—but at a civilized Chicago café? Incredible. Harry James tried to put the idea out of his head—
but instead of disappearing, it grew more obsessive.
How would reckless, fast-moving Chicago society take to
such an audacious experiment? James was a perceptive
student of human nature—you had to be, in this business—and he thought he knew. So in February of 1916
he wired Johnny Stein to bring his band to the Booster's
Club, a ten-week contract guaranteed.

Stein stared unbelievingly at the telegram. No one
in the band had taken James seriously that night back
in December. He hurried over to Magazine Street immediately to tell LaRocca, and the two of them discussed
plans for their northward expedition. LaRocca succeeded in convincing Stein that Eddie Edwards was the
trombonist to use in place of Leonce Mello. Mello, he
said, played "big cornet"—too much melody and not
enough counterpoint. Edwards was actually three trombonists in one—he played harmony part of the time,
counterpoint part of the time, and at other times accented
the beat like a bass drum. With Edwards a band always
seemed larger than it really was, for he gave it body.
In the words of LaRocca sometime later, "It's like a
dress—I cut the material, Shields puts on the lace, and
Edwards sews it up."

The decision to go north with the band was not an
easy one for Edwards to make. He had been playing
semiprofessional baseball around New Orleans and now
had an opportunity to join the Cotton States League at
Hattiesburg, Mississippi, as a third baseman. Baseball

was a strong love, but music was stronger. Eddie joined the band.

LaRocca looked around for Larry Shields, but the adventurous clarinetist had gone to Chicago a whole year earlier with Tom Brown's ragtime band, unknown to his hometown friends. So Alcide "Yellow" Nunez—so named because of his peculiar complexion—stayed with them on clarinet. "Yellow" was known all over New Orleans as the only man who could take his clarinet to pieces down to the mouthpiece and still keep up with the band. But LaRocca always preferred Shields to Nunez because of the latter's stubborn insistence on playing melody. The famous three-voiced interplay of the dixieland ensemble was always prominent in the mind of LaRocca and he fought constantly for a clear field on the melody. At this time the distinctive brand of counterpoint that was to be a distinguishing feature of jazz was unfamiliar to New Orleans musicians.

After two short weeks of rehearsals at Johnny Stein's house, the band, now composed of Stein, La-Rocca, Edwards, Nunez, and Ragas, proclaimed itself ready for the test, and on the morning of March 1, 1916, departed for Chicago.[2]

This writer talked with Harry James at his home in New Orleans in May of 1957. Then in an extremely critical state of health and under the care of his daughter, he had lost none of his enthusiasm for the band he discovered. Vividly he described his reactions to the first

[2] LaRocca had been hired by Jack Laine to play in the first carnival parade of the Mardi Gras on March 2, 1916, but Nick sent a trumpet player named John Provenzano to take his place. Laine didn't know until several days later that Stein's band had gone to Chicago.

crude attempts at jazz, the thrill of bringing Stein's band —the embryo of the Original Dixieland Jazz Band—to Chicago, and his paternal devotion to the green young musicians from New Orleans. James's success as a café manager continued in Chicago for a few years until the Original Dixieland Band left for England in 1919. Then he began his wanderings, serving as steward on an ocean liner, as roustabout, as bartender in miscellaneous dives from Cairo to Calcutta, and finally as assistant manager of the Bentley Hotel in Alexandria, Egypt. In 1924 he returned to the café business in New Orleans. As late as the summer of 1940 he was managing Gasper's 440 Bar on Bourbon Street. In a way, Harry James could be called a chronic adventurer, a soldier of fortune. He pursued his life with a furious passion and spent it like a spendthrift. But if he accomplished nothing else, he was, according to Paul Whiteman, directly responsible for the jazz craze that swept a world.

3

Chicago: The Cradle of Jass

The exploration of Chicago by early New Orleans brass band musicians began as a trickle of adventurous souls and gradually developed into a mass migration of hungry opportunists. It is difficult to say just when the first brass ragtime band sounded its blaring notes in the Windy City, but among the first was Tom "Red" Brown and his squad of travel-worn minstrels, who arrived there in the spring of 1915. The six-piece combination was brought to Chicago by Louis Josephs, more popularly known by the stage name "Joe Frisco."

Joe Frisco was one of the most picturesque vaudeville entertainers of the period. The whimsical, stuttering, rubber-legged little man with the derby hat was literally born for the stage, for his every gesture was conducive to laughter. His specialty was an act called "A Charlie Chaplin Imitation." Dressed in the familiar black derby, diminutive moustache, and coat and tails of the famous silent movie comedian, he waddled about the stage and danced an original soft-shoe routine. As he skated and slid about the floor, rolling his derby up and down his arm and twirling his cane, his audience was driven into paroxysms of laughter.

One night in November, 1914, Joe Frisco was doing his act at the Young Men's Gymnastic Club in New

Orleans. Furnishing the music was Tom Brown's band. Frisco was obviously impressed by his accompaniment, because six months later, when he was preparing a show for Lamb's Café in Chicago, he remembered Brown's aggregation and brought them north for a six weeks' engagement, beginning May 15, 1915. The band at that time was composed of Ray Lopez (cornet), Tom Brown (trombone), Gus Mueller (clarinet), Arnold Loyacano (guitar), Steve Brown (bass viol), and Bill Lambert (drums). They functioned without a piano, the guitar and bass viol providing the chord foundation. Billed as "Brown's Band from Dixie Land," their music consisted mostly of fast ragtime and wild renditions of "Turkey in the Straw," "Listen to the Mocking Bird," "Reuben Reuben," and similar tunes.

It was during the Lamb's Café engagement that bandleader Bert Kelly asked Tom Brown where he could find a good clarinetist. Brown recommended Larry Shields and immediately sent a telegram to New Orleans. Shields rushed to Chicago with all the anxiety of a kid on his way to a circus and joined Kelly's band at the White City. But the northern musicians of the White City outfit confused Larry, who couldn't quite figure out how to read music, and the young clarinetist became morose and melancholy in his new job. Tom recalls the night he saw Larry in the corridor of the Chicago hotel where they were both staying. Despondent, Larry had just come off the job with Bert Kelly and was complaining desperately that he couldn't play the music. "I just don't fit," he said sadly. An electric light flashed in Tom Brown's head. His own clarinetist, Gus Mueller, had been restless and inclined toward moving on to another band. So Brown immediately arranged the swap, and on August 4, 1915, Larry Shields became a member of Brown's Band from Dixie Land.

Despite their importance in the formative period of the new music, the history of Brown's band reads like a textbook of failures. The attendance at Lamb's Café dwindled steadily. Finally, on August 28, 1915, when the place closed for repairs for three weeks, Brown's band headed east on a vaudeville tour, calling themselves, "The Kings of Ragtime." In December booking agent Harry Fitzgerald got them a job at New York's Century Theater in Ned Wayburn's *Town Topics* review. The show had been running more than a month when Brown's musicians joined the cast, but it folded six days later. Brown claims they hung around the big city for eleven weeks, drawing pay under the terms of their contract. Early in 1916 they went into vaudeville again, billed as "The Ragtime Rubes." In this act the musicians dressed in overalls, bright-colored shirts and straw hats, and Larry Shields played the part of "the silly kid."

They played the Columbia Theater where, according to Tom Brown, "the applause was like rain." But the rain must have ended in a complete washout, for the group disbanded in February, 1916, and retreated to New Orleans. Brown maintains that pay was not immediately forthcoming from the management and that Larry Shields, hungry and disappointed, quit in a mood of utter despair, ruining the act.

There was, therefore, no other New Orleans band

in Chicago when Stein's band arrived there on March 3, 1916, most of the southern musicians having been driven back home by cold winds and empty stomachs. It had been a long, hard winter in the Windy City, and the frigid gales continued mercilessly into the early spring.

Stein's party of five had never been north before and were understandably ill-equipped for their new experience. Harry James, seeking to protect his investment from multiple pneumonia, quickly herded them into a secondhand clothing store where he bought each of them a long overcoat of incredible weight. "They had never seen overcoats before," recalls the dapper, well-dressed Harry James with a chuckle, "and when they walked out of the store in these long, black overcoats, they looked just like five undertakers!"

Unfortunately, the Booster's Club, for which they had been hired, had been closed by police order a few days before they arrived; but through the further efforts of Harry James they were secured an audition at Schiller's Café on the south side of the city. Ten people were present at the rehearsal, and LaRocca recalls the plight of the five frightened musicians on that memorable night. Failure would have meant a complete fiasco for them in the North and a humiliating retreat to New Orleans, where most people had doubted the success of their Chicago venture.

Their music went over. By eleven o'clock on Saturday, the opening night, the club was filled to capacity and crowds milled about the entrance. Firemen were dispatched to control the throng, allowing two people admission for every ten leaving. They were soon the talk of

Chicago night life, and these clean-cut southern lads who called themselves simply "Stein's Band from Dixie" had made Schiller's Café on 31st Street a most popular spot.

It was during their run at Schiller's that the word "jass" was first applied to music. A retired vaudeville entertainer, somewhat titilated by straight blended whiskey and inspired by the throbbing tempos of this lively band, stood at his table and shouted, "Jass it up, boys!"

"Jass," in the licentious slang vocabulary of the vast Chicago underworld, was an obscene word but like many four-letter words of its genre, it had been applied to almost anything and everything and had become so broad in its usage that the exact meaning had become obscure to most people.

Harry James, now the manager of Schiller's, never missed a bet. When the unidentified inebriate bellowed forth his now-famous "Jass it up, boys!" (and by so doing, unwittingly wrote a full page of musical history), the gold-plated thinking machinery of the Chicago café expert was once more set in motion. The tipsy vaudevillian was hired to sit at his table and shout "Jass it up," every time he felt like it—all drinks on the house. The next day the band was billed, in blazing red letters across the front of Schiller's:

STEIN'S DIXIE JASS BAND

Chicagoans then had a word for the heretofore unnamed music.

One important fact deserves emphasis: The dixieland band that came from New Orleans was a group of self-taught musicians still experimenting with technique. The music they played bore little similarity to the music for which they are now noted. They lacked skill and had not yet formulated a special kind of music. It was always plentiful in volume, but the New Orleans band played in a slower tempo. The clarinetist had not acquired sufficient skill for improvising continually around the melody, and he usually played in unison with the cornet, or in harmony when he knew how. The trombone carried a simple counterpart with loud blats and long glissandi.

During the months of daily practice and nightly performance in Chicago, an important evolution took place. The artists finally mastered their medium. More sure of themselves, they speeded up the tempo to satisfy the frenzied spirit of the dancers. LaRocca's cornet devised a new, driving style; the clarinet began its first awkward attempts at "noodling" around the melody; Edwards' trombone accented the beat with deep, powerful tones and overlapped LaRocca's phrasing with a new type of counter-melody—a style partly resembling the old military bands of New Orleans. The strong trombone was a very important feature of this new music; in the words of LaRocca, "Daddy Edwards was the daddy of all tailgate trombonists."

This, then, was the music to which the word "jass" was first given. The outstanding feature was its counterpoint. The three parts fitted together like the pieces of a jigsaw puzzle. The cornet played melody, left gaps to be filled in by the clarinet, and was complemented by the trombone. The big parade drums provided a firm foundation, and the rumbling "back room" piano added the finishing touches. It was a fast, traveling, two-beat style.[1]

This was "jass" as Chicago knew it. It was ensemble playing with contrasting strains. There were no solos. For the most part, the tunes played were rags, and *it was the instrumental arrangement as much as the new rhythm that distinguished this music as "jass."* There were changes in rhythm and accent as the tempo was increased, but these were secondary characteristics to the band's radical ensemble effect.

An ironic feature of any real jazz band is that regardless of who organizes, manages, or does the talking for the outfit, it is always the cornet (or trumpet) player who is the real leader of the band. The very musical

[1] The question of whether the rhythm of the Original Dixieland Jazz Band was two-beat or four-beat is deceptively simple on the surface. Actually, even the original members themselves cannot agree. LaRocca and Sbarbaro are sure that it is two-beat, because of its parade-band origin; Edwards insists that it is four. The author is inclined to go along with Edwards and has always believed that the Dixieland Band's fast one-steps had a four-beat "feel." Unfortunately the recordings of the band were not at all typical of their output and are misleading as examples. Don Fowler ventures the opinion that it is neither: "The salient element is syncopation. Dixieland tends to accent beats 2, 3, and 4. It hurries the beat, plays ahead of it. In the original records, Sbarbaro does not play two-beat or four-beat, but whatever fits the tune at a certain moment. The original dixieland was syncopated, contrapuntally conceived. The element of rhythmic surprise was certainly a factor." And J. S. Moynahan agrees: "The question of 2-beat or 4-beat is one of those tempests in a teapot which amateurs are always raising. It is a combination. Tony has a neat trick where he plays the bass drum after the beat a couple of measures and then returns to the beat. Let's say that it was more two-beat than is generally played today, but that it varied constantly with syncopation and musical emphasis." But perhaps the statement of Santo Pecora, whose dixieland band is currently playing on Bourbon Street, is the clincher: "We used to play two-beat but had to switch to four to keep the customers from walking out. It's the four-beat that gets 'em."

structure of the band makes it so. The cornetist, as leading voice in the ensemble, "calls the shots"—starts,
stops, and determines the melody and even the key it is
played in. The cornetist is the captain of any jazz band,
for better or for worse, and the entire aggregation rides
his ship.

Nick LaRocca, therefore, was always musical
leader of the bands in which he played—Jack Laine's,
Stein's, and the subsequent Original Dixieland. In this
capacity he was unrivaled. No one ever "drove" a band
like LaRocca, and to this day his secrets of "drive"—
the tricks of "blowing in" a phrase, of hitting slightly
ahead of the beat, of dropping out at a critical moment
—have never been equalled. The well-known music
critic Charles Edward Smith was one of the first to express this in words. Writing in the British magazine
Swing Music during the early 'forties, Smith says:

... LaRocca experimented with his horn ... The edge of his
tone was clear, the notes cut off sharply to permit of succinct
phrasing. Some of his breaks were elementary, others, like the
"flying tackle" and the "ragtime shuttle," have never been
duplicated. He also had a peculiar trick of "blowing in" a
phrase with a stimulating volley of sound. Aside from these
physical contours, LaRocca's cornet style had body; if one
fails to remark the rightness of his slightly bland tone it is because it is so perfectly in keeping with the band as a whole.

Whatever the characteristics of LaRocca's style,
they were the product of his unique self-education. Some
evolved out of what an academic, conservatory-trained
musician would call "bad habits"; others were the results of his curiosity and boundless imagination. Above

all, the relentless drive exhibited in LaRocca's style was the very reflection of his personality, the ultimate integration of man and instrument.

The desire to lead, to have things all his own way, to take command of every instrument was always present in Nick LaRocca. Tucked away in the cupboards of his turbulent musical mind were a dozen original compositions, all waiting for the opportune moment. They were being saved for the day he would have his own band, for these compositions were not just melodies—they were whole arrangements for cornet, clarinet, trombone and rhythmic background, and they demanded control over every musician. The kind of counterpoint that later distinguished the Original Dixieland Jazz Band was born in his musical imagination. The "conversations," as he preferred to call them, between the various musical instruments—the shouting, answering, arguing, and laughing of cornet, clarinet, and trombone were the indirect fruits of his adolescent experience, the nights at the French Opera when he listened in awe to the fugues or "conversations" of Italian opera singers.

These spirited jazz creations were like caged animals within his brain, all fighting to break loose. One of them, "Livery Stable Blues," succeeded under the following very curious circumstances. The band had been playing two weeks at Schiller's and the enthusiastic patrons had been growing increasingly boisterous and hysterical over the music. Edwards describes the occasion:

There were several people in the Schiller's Cafe and one girl in particular was evidently feeling jolly and sky-larking to the

amusement of the boys in the band, which prompted LaRocca
to pick up the cornet and play a horse whine on it. Everybody
laughed within hearing distance of it, and I told him at the
time it would be a good stunt to put this horse whine in a
number, and he said he had it in a number. I, of course, asked
him what number it was, and he replied, "a blue number." I
told him some time we might try it and it might prove a good
number to us.[2]

In the later morning hours, after closing time, the
band rehearsed the new "Livery Stable Blues," LaRocca
humming the various parts to the other musicians, or
picking them out on the piano. The horse's neigh, or
"whinny," had been discovered by accident many years
earlier when a valve stuck on his cornet. Experimenting
further, he had found that by holding down the third
valve and shaking the horn, a lifelike imitation of this
"whinny" could be produced. Now he instructed the
clarinetist to emit a rooster crow and the trombonist a
donkey bray. These three sound effects comprised the
"breaks" in "Livery Stable Blues." The chord structure
was borrowed from Stephen Adams' "The Holy City"
and modified by changing one chord and deleting the
last two measures. The melody imposed on this foundation was completely original.

In the lawsuit of October, 1917, the late Henry
Ragas describes the rehearsal: "He [LaRocca] said he
will play an old number of his, and he said, 'I will go
over it and give you an idea,' and he played it over on

---

[2] From the court records of the "Livery Stable Blues" law-
suit, District Court, Northern District of Illinois, Eastern Division.
Testimony of Edwin B. Edwards, September 28, 1917, at office of
Rae Hartman, 165 Broadway, New York, N. Y. Signed copies in
LaRocca Collection.

the cornet softly and told me just exactly when to stop for the runs to come in." [3]

After the immediate success of "Livery Stable Blues," three more of LaRocca's original compositions, "Tiger Rag," "Sensation Rag," and "Ostrich Walk," were rehearsed and developed in that order. (The titles emphasize the composer's obsession with animal sounds, as many of his tunes were mentally pictured as beasts. The musician who plays by ear, and who has never read music, visualizes music in terms of images or abstract patterns, rather than marks on a musical staff.) The new novelty numbers, especially "Tiger Rag" with its roar of the tiger, executed on trombone, increased the fame of the "jass" band, and business hummed at Schiller's.

For three months, Stein's Dixie Jass Band held forth at the popular 31st Street café, where nightclubgoers fought one another for admission, liquor was served quite illegally until daybreak, and drunken orgies ran unabated. On Saturday night, April 29, the Anti-Saloon League, in co-operation with the Church Federation, sent a delegation of sixty women on tour of the southside cafés, apparently to investigate certain immoral and illegal aspects of these places. Quite inevitably they found their way to Schiller's, and the results of their "investigation" were published in the Chicago *Herald* for May 1, 1916, providing an excellent first-hand account of the environment in which "jass" was fostered:

[3] From the court records of the "Livery Stable Blues" lawsuit. Testimony of Henry W. Ragas, *ibid.*

SIXTY WOMEN RIP MASK FROM VICE

. . . The party . . . adjourned to the New Schiller Cafe, at 318 East Thirty-First Street.

A line of taxi cabs radiated from the Schiller to the east, west, north, and south. In front of the doors, a crowd of people fought for admission. A perspiring doorman held them back. "Can't come in," he shouted. "We're crowded to capacity. Wait 'till some of the others come out."

This was 2:30 o'clock in the morning.

The crowd in front of the doors kept increasing all the time and the doorman had his hands full keeping the mob from rushing him off his feet. No policeman was in sight. The party finally obtained admittance and a table after much elbowing and shoving. It was impossible for anyone to be heard. The shriek of women's drunken laughter rivaled the blatant scream of the imported New Orleans Jass Band, which never seemed to stop playing. Men and women sat, arms about each other, singing, shouting, making the night hideous, while their unfortunate (?) brethren and sisters fought in vain to join them.

The party ordered gin fizzes, cocktails and beer. They were served in a jiffy . . .

LaRocca takes issue with the reporters:

The part about no policeman being in sight was wrong. The Fire Department had the management cut exits through the back of the building, and they did have policemen and firemen stationed at the Schiller Cafe. . . . After the sensation we created, other cafe owners sent to New Orleans for men who were supposed to play our kind of music. They imported anybody that could blow an instrument, and they all had "New Orleans Jass Band" in front of their places. . . . We were never advertised in the papers, as every night was a full house. . . . The impact we had on the people of Chicago was terrific. Women stood up on the dance floor, doing wild dances. They had to pull them off the floor. The more they would carry on, the better we could play. Here is where the singers came to the rescue, as these patrons would never leave the floor and the manager wanted them to sit and drink. Then the crowd would start yelling, "Give us some more jass." I can still see these women who would try and put on a show dance, raise their dresses above their knees and carry on, men shrieking and everybody having a good time. I would let go a horse whinny on my cornet and the house would go wild.[4]

Despite the prosperity of Schiller's Café, no salary increases were forthcoming for Stein's band, and the members found it increasingly difficult to live on twenty-five dollars a week. They needed new clothes, especially uniforms to replace the "dusters" they wore as substitutes, and extra cash for spending money. LaRocca and Edwards convinced Nunez and Ragas that the time had come to move. By the unanimous consent of these four musicians it was decided to forsake Schiller's in favor of a higher-salaried engagement elsewhere in town. Stein refused to take a chance on breaking the contract, an argument ensued, Edwards punched Stein in the nose, and on May 26, 1916, the rebellious four took their leave of Schiller's Café in an explosion of southern accents.

Then it was that the Original Dixie Land Jass Band came into being.

[4] From a letter to the author, dated March 16, 1959.

4

Jass and the Underworld

Sammy Hare, the owner of Schiller's Café, flew into a rage. This was gratitude for you, he thought. Give these five kids a break, they make good, and what happens? They leave you holding the bag. With a trembling hand, he picked up his telephone and called Harry James. He pleaded and bargained, ranted and raved, but the sober café manager was unmoved. "What can you expect?" said James, fingering his diamond cuff links. "Everybody has to eat, you know."

Hare served papers on the four deserting musicians, charging breach of contract. But the New Orleans boys were not without friends in their predicament. Sam Rothschild, the affable manager of Del'Abe's Café in the Hotel Normandy, had been watching the developments at Schiller's with a great deal of interest. A devout fan of the new "jass" music, he often came to the rival café nightly to listen and to chat with the musicians. He thought the Dixie Jass Band was being unfairly exploited, resented the dealings of his competitor and arch enemy, and saw the chance to play his cards. He had encouraged the "jass" band to leave Schiller's; now he offered them the services of his lawyer, Timothy J. Fell.

The rebellious four called upon Timothy Fell and explained their case. The lawyer settled the matter for them without leaving his swivel chair. He said that Stein, as leader of the band, was solely responsible. But he advised them to appeal for an earlier hearing and then told LaRocca exactly what to tell the judge, rehearsing him word for word. The judge must have been strongly impressed by the rustic simplicity of these four poorly dressed musicians from New Orleans, with their gentle southern accents. He freed them from their contract immediately after LaRocca pleaded that they were not receiving a living wage at Schiller's, and that they would become a charge on the city of Chicago if they were denied the right to perform in local cafés. The case could not have been presented more effectively by the shrewd Timothy J. Fell himself.

Meanwhile, news of the overwhelming popularity and success of the Dixie Jass Band had spread to New Orleans. Suddenly the northward migration of parade band musicians turned into a wholesale excursion. The trains running between the Crescent City and Chicago became a shuttle service for ambitious musicians, eager groups venturing northward, who passed disappointed, starving compatriots on their way back home. For none of them, try as they might, could divine the secret of the LaRocca-Edwards combination; none of them could supply the new "jass" music that Chicago night life clamored for.

Johnny Stein finished out his contract at Schiller's with a band composed of "Doc" Behrenson (cornet), Jules Cassard (trombone), Larry Shields (clarinet), Ernie Erdman (the house pianist at Schiller's), and himself on drums. At the conclusion of the Schiller job, sometime in May of that year, Stein left for New York

City. In the early twenties he was at Coney Island, playing drums in Jimmy Durante's jazz band.

With their litigation settled, the four jazzmen who
seemed to be on top in the always-competitive music
world—LaRocca, Edwards, Nunez, and Ragas—began
rehearsing their new combination, a co-operative organization to be known as "The Original Dixie Land Jass
Band." LaRocca was elected the musical leader (and it
is reasonable to assume that some active campaign ora-
tory on his part preceded the election), Edwards became
the business manager, and the two of them functioned
for awhile as sort of co-captains.

On Friday evening, June 2, 1916, they opened at
the Del'Abe Café in the Hotel Normandy, at Clark and
Randolph streets. They brought in customers by the
dozen as Sam Rothschild stood smiling in the doorway.
They sent for a New Orleans drummer by the name of
Anthony Sbarbaro, and Earl Carter of Chicago filled in
for two weeks pending his arrival by train.

The Dixieland Band was gradually approaching the
final stages of its evolution, and each new change brought
it just that much closer to perfection. The latest improvement was Tony Sbarbaro. Born June 27, 1897, he was the
youngest member of the band, not having quite reached
his nineteenth birthday when he arrived, wide-eyed, in
Chicago. He had been playing at the Tango Palace in
New Orleans with the ubiquitous Brunies brothers, and
a year earlier had been a member of a band led by
Ernest Giardina which included cornetist Emile Christian and trombonist Eddie Edwards.

In the opinion of many experts, Tony Sbarbaro was
one of the greatest of all dixieland drummers. In keep-

ing with the rest of the band, he was traditionally inventive. He was the first drummer to use cowbells; and his famous kazoo, which he picked up in a novelty store in Chicago, immediately set him apart from all others of the period. The collection of dolls, teddy bears, and miscellaneous gimcracks with which he decorated his drum installation established him as a forerunner of the modern "hot-rodder." Possibly his hoard supplies a clue to the origin of the phrase "trap drums," as anyone falling headfirst into this assortment of junk in the dark would certainly consider it a trap.

The drumming of Tony Sbarbaro has been referred to by his contemporary descendants as a "gallopin'" style, and this seems to describe it better than a whole paragraph of technical terms. Basically, it was a three-stroke "ruff" with an accented press roll. Again, the early New Orleans parade bands showed their influence. Tony used a gigantic, twenty-eight-inch street drum and made himself heard blocks away. At least one and preferably two cowbells were essential to the new style, plus a large-sized woodblock for continuous use during the choruses, when it replaced the snare drum. Sbarbaro's practice involved use of the snare on verses, "minstrel style" woodblock on choruses, returning again to the snare drum on the last half of the last chorus. Naturally, brushes were never used. The "minstrel" woodblock reveals the influence of the minstrel shows, so common in New Orleans at that time.

The success of the Original Dixie Land Jass Band reached new heights at Del'Abe's, when hundreds of their fanatic fans from Schiller's came over to augment the growing café clientele from Hotel Normandy. Meanwhile, the ever-industrious Harry James had bought a half interest in the Casino Gardens at Kinzie and North Clark streets, just outside the Loop, and now persuaded the band to move there for a long-term engagement beginning Thursday, July 6, 1916.

Business boomed in a similar manner at the Casino. James had the walls torn down and the dining room enlarged to accommodate the increased demand for tables. While employed at the Casino, the Dixie Land Jass Band also tried its hand at vaudeville. The effect of "jass" upon uninitiated Chicago theatergoers is colorfully illustrated by a press review in *Vaudeville*, August 31, 1916:

. . . Fogarty's Dance Revue and Jass Band hit it off like a whirlwind in next place. The Jass band was a hit from the start and offered the wildest kind of music ever heard outside of a Commanche massacre. There are five men in this band, and they make enough noise to satisfy even a north side bunch out for entertainment in the vicinity of Belmont Avenue and Racine . . .

Said the Chicago *Breeze*, September 1, 1916:

"Jimmy Fogarty's Dancing Revue and Jass Band" is a new act which is seen at McVicker's this week. Johnny Fogarty, who has been prominent as a dancer for many years, and who has had big acts with society dancers for the last few years, has seen the popularity of "The Jass Band" and turned it into vaudeville. He has three couples for dancing, and the band not only accompanies most of these exhibitions, but it has a number to itself. The band is composed of five white men—piano, drums, slide trombone, cornet, and clarinet . . . The band took big applause Monday night at the first show (17 minutes, full stage).

And in *Billboard*, September 1, 1916:

No. 5—Fogarty's Dance Revue and Jass Band, consisting of eight men and three women offered a conglomeration of modern dancing and incidentally introduces vaudeville's newest craze, "The Jass Band," and the way the five of them tear away at their instruments brought down the house. Each member is an artist. The revue is elegantly dressed and all step lively to the tuneful sounds of the band. Everyone co-operated in its success. Twenty-three minutes, full stage; two curtains.

Although the Dixie Land Jass Band was beginning to click, there remained one weak spot in its constitution. The unreliability of "Yellow" Nunez became a constant source of worry to the more serious members of the band. In its incipient stages, this lack of interest merely took the form of tardiness at the bandstand. The band would be poised on the stand, ready to begin a set, while Yellow would be talking with patrons in a darkened corner of the club, or getting himself lubricated at the bar upstairs. His delinquency increased at an alarming rate; sometimes he would wander off the bandstand and disappear for the balance of the night. At other times he wouldn't even show up for the first number. In this case the band would form a posse, making the rounds of likely neighborhood spots in search of the missing clarinetist. The drinks he charged to the band reached staggering totals, and the payment of the musicians became confused.[1] On October 31, 1916, La-

---

[1] Eddie Edwards, as business manager, claims that he would pay Nunez in one-dollar bills and give him a "short count" (counting the same bill twice, etc.) each week until Nunez' debt was paid off. LaRocca, on the other hand, refutes this statement with the explanation that Edwards did not handle the money for the band, and that in Chicago the members were always paid individually by the management.

Rocca fired Nunez from the Original Dixie Land Jass Band. The combination limped along without a clarinetist for the next three days, pending the arrival of Larry Shields.

Nunez returned to New Orleans, where he formed a band with Emile Christian (cornet), Sigmund Behrenson (trombone), himself (clarinet), Eddie Shields— brother of Larry (piano), and Johnny Stein (drums). The group ventured to the Vernon Café in Chicago where they billed themselves as a "jass" band with very little success. About a year after the Original Dixieland Band had made their first phonograph record, Nunez put together a band called the Louisiana Five and made one side of a record for Columbia (No. A2768, serial No. 78523), very appropriately titled "The Alcoholic Blues." Readers who are curious about the style of Alcide Nunez will learn much from this rather obscure disc, which features a combination of clarinet, trombone, guitar, piano, and drums—no cornet. Nunez leads the band on clarinet, playing straight melody from start to finish without so much as a single improvisation. (This bears out LaRocca's claim that Nunez always fought with him for the melody, resulting in an untuneful clash between cornet and clarinet—or in LaRocca's term, "bucksawing.") His tone is strong, broad, and reminiscent of the old-time parade band musicians. In 1927 he returned to New Orleans, where he joined the police force, and in his spare time played in the New Orleans Police Band.

Larry Shields was born September 13, 1893, and first showed an interest in music at the age of fourteen, when he bought a secondhand clarinet from a New Orleans musician of local fame. Like LaRocca he was entirely self-taught, could not read music, and learned everything by trial and error. This writer is not alone in his opinion that Shields was the greatest jazz clarinetist of all time. His technique was flawless, his tone clear and full of character, and his ideas twenty years ahead of their time. The "noodling" style of clarinet solo so successfully exploited by Benny Goodman during the "swing" craze of 1936–38 had its seeds in the ensemble playing of Larry Shields and was well displayed in the Original Dixieland Band records of 1917 and 1918. Shields was, in fact, long the idol of Goodman, who practiced at an early age by playing along with these records.

The addition of Larry Shields brought the Original Dixie Land Jass Band to its final configuration. It was this combination of LaRocca, Edwards, Shields, Ragas, and Sbarbaro that went on to the heights of popularity reached by no other jazz band of that period.

By 1916 the growth of gang warfare in Chicago was well under way. Two or three bodies were fished out of the Chicago River every week, and the "protection" racket was becoming one of the city's leading industries. Although the "Tommy gun" and the long black touring car had not yet come upon the scene, the ruthless gangsters of the Windy City managed very well with what they had.

The friendliness and good-natured innocence of the five Southern gentlemen were no assets in this seething metropolis. In those days it was hard to remain neutral. Before you knew it, your best friend was connected with some gang, and you needed the gang for protection.

The police, corrupt from the "beat" to the top desk, conveniently looked the other way or just weren't around during troublesome outbreaks.

Unknown to Harry James, making its headquarters at the Casino Gardens was the North Side Gang of Mickey Collins. By night, this mob of underworld characters systematically pillaged the surrounding communities but, as if some code of ethics was involved, they never robbed within a fifty-mile radius of the Casino. In the afternoon they returned to the café with their loot, called the roll, and took inventory. LaRocca recalls their flashy moll, a slim brunette named Faye, who arrived every day loaded with jewelry and trinkets.

In the midst of all this opulence, the plain, bargain-basement attire of the band members must have seemed appalling to the showy mobsters. The poorly fitting black overcoats donated by Harry James had been abandoned the first time the musicians were called "hayseeds" by wisecracking Chicagoans. Now they wore inexpensive clothes which fitted but which were woefully inadequate for the heavy northern winters. During the coldest part of December they stuffed newspapers down the backs of their flimsy topcoats to protect against the biting Chicago gales.

One afternoon as LaRocca was hanging up his coat and removing the wrinkled newspapers, one of the "boys" known as Earl Deer swaggered over to him and asked him why he didn't buy himself a decent overcoat. When Nick explained that he had been sending most of his money home to his mother and couldn't afford it, Deer said, "Meet me at Sears Roebuck tomorrow afternoon and I'll buy you one."

Nick waited at the store that day for an hour, trying on overcoats and expressing his choice to the clerk,
but Deer never showed up. When Nick reached for his
old topcoat that night at the Casino, he found that it had
been slashed to ribbons. The next afternoon Deer presented LaRocca with the very same overcoat he had
picked out at the clothing store the day before—or one
very similar to it. The wisdom of accepting a gift from
the mob was dubious, but Nick, with his old coat torn
to shreds, was afraid to return the new one and too cold
to give it away.

From that day on, the musicians decided they
would have nothing more to do with the Collins' mob.
But such hasty resolutions were easily forgotten by the
impetuous young bandleader, whose mind was on women
more often than music. Nick had been dating an athletic
blonde by the name of Jessie, a popular lady wrestler in
those parts. But unknown to him, Jessie was the girl-
friend of Joe Bova, leader of the rival South Side Gang.
Bova was quick to make his displeasure known to the
cornetist—so unmistakably, in fact, that LaRocca began
carrying an automatic in his coat pocket and was never
on the bandstand without it. "We called him 'Nick the
Gunman,'" says Edwards.

One night, after Nick had defied the notorious
gangster by spending that afternoon with Bova's lady-
friend, he was paid a visit by Bova. Bova, without
mincing any words, told LaRocca that if he wasn't out
of town by sunrise the Chicago River would claim a
waterlogged cornet player.

Mickey Collins had been watching this conference

from across the room. When it was over he hastened to the bandstand.

"What's that punk doing in here?" demanded Collins, blowing a cloud of blue smoke through his nose. He listened carefully while LaRocca explained his predicament. A few minutes later Bova was called upstairs on the pretense of a telephone call, beaten to a pulp by Collins and Deer, and pinned to the wall. (The custom of having one's ears pinned to the wall, usually with long hatpins, was as common as the flu in Chicago at that time.)

Despite the undercurrent of gangland activities, the Casino Gardens grew in popularity as one of Chicago's gayest night spots. Billed as "The Theatrical Profession's Most Popular Rendezvous," it was soon the accepted meeting place of transient actors, dancers, singers, and musicians. Through the public relations genius of Harry James, this place became the headquarters of the Theatrical Boosters' Club. Such dignified guests as Will Rogers, Fannie Brice, and Bert Williams stopped by regularly to dine and listen to the exhilarating rhythms of the "jass" band, which by now was becoming known far and wide in show business.

The up-and-coming Al Jolson made a special point of stopping in to see the five-piece band he had heard so much about from his fellow performers. According to Harry James, the highly emotional Jolson was moved to the point of tears the first time he heard their music. They commanded his attention to such an extent that his dinner went cold and uneaten, and members of his party were totally ignored for the duration of the musical act. When Jolson buttonholed the well-known theatrical agent, Max Hart, in New York weeks later, he did such a thorough job of selling that Hart made a special trip to Chicago just to hear the Original Dixie Land Jass Band. When he had heard them, he signed them immediately for a two-week try-out at Reisenweber's Restaurant in New York City, to be followed by a guaranteed contract of $750 per week if the band made good.

The popular movement which Nick LaRocca later referred to as "The Revolution in Four-Four Time" was about to begin.

New York and the Jass Revolution

It was January, 1917. Germany was preparing to resume unrestricted submarine warfare on February 1,
and most Americans still did not suspect that the
United States would be drawn into the conflict within
three months.

In New York City, people tried to forget international troubles. Some went to see Eva Tanguay and the
Marx Brothers in the vaudeville at the Palace. Others
saw Nora Bayes at the Eltinge, or stood in line for
choice seats at *Treasure Island,* the season's smash hit.
College boys flocked to see Anna Held in "the girliest
show in town" at the Casino.

The January 15 issue of the *New York Times* carried a full-page advertisement by Gimbel's Department
Store, announcing a big sale of "fur-lined coats for
motoring." The largest ad on the amusement page read
as follows:

Margaret Hawkesworth's
"PARADISE"
The Smartest, Most Beautiful
and Most Modern Ballroom in America
in the Reisenweber Bldg
at Eighth Avenue and 58th Street

announces

The First Sensational Amusement Novelty of 1917
"THE JASZ BAND"

Direct from its amazing success in Chicago,
where it has given modern dancing new life and a new thrill.
The Jasz Band is the latest craze that is sweeping the nation like a musical thunderstorm.

"THE JASZ BAND"

comes exclusively to "Paradise," First of all New York
Ballrooms,
and will open for a run TONIGHT (Monday).
You've Just Got to Dance When You Hear It.

The debut of the Original Dixieland "Jasz" Band
at the Paradise was an experiment that was very cau-
tiously undertaken by the management. The band began
playing there on January 15 for a two-week trial, during
which they played two numbers each night while the
regular house orchestra rested. The response at first was
disappointing. Blasé New Yorkers vacated the dance
floor and stared numbly while the southerners knocked
themselves out with their own fast, blaring, syncopated
music. Most of the spectators considered the whole
affair a rather audacious publicity scheme.

On Wednesday, January 17, an advertisement in
the *New York Times* announced the formal opening of
the new Reisenweber Building, "New York's newest,
largest and best-equipped restaurant with private dining
rooms, ball rooms, beefsteak grill, tiled kitchens, ty-
phoon ventilation, and seven dance floors." Gus Edwards' girlie revue, Round the Circle, with Norton Lee
and a company of thirty entertained guests in the Main
Dining Room, while in the Paradise Ballroom, Mar-

garet Hawkesworth and Alexander Kiam performed exhibition dances. But what had happened to that "sensational amusement novelty of 1917," the "jasz" band? The habitual reader of the *New York Times* amusement page might have had cause to wonder.

The public's initial apathy for the new music, demonstrated in the Paradise on Monday night, had temporarily discouraged the management. But two numbers had not been a fair trial for any such radical experiment, and Monday night was not the most popular night in the week for dancing. Public response to "jasz" increased rapidly during that critical week. After the first week end the management decided to hold the band for the formal opening of the "400" Club Room on Saturday, January 27. The "400" Room, originally scheduled to open with the rest of the new building on the seventeenth, had not been ready in time because of delays in interior decorating. In the meantime the five musicians were busy getting fitted for their first tuxedos (they owned no uniforms), but found time to play another "experimental" set at the Coconut Grove.

When the band arrived at the "400" Room on opening night they found themselves in total darkness. Feeling their way about the place, they located the bandstand, helped Tony set up the drums, and then the two ex-electricians, LaRocca and Edwards, went into the basement to trace the house wiring. After some preliminary investigation, LaRocca discovered that the wiring had been improperly connected to the main fuse box, and after reversing the polarity of the lines, quickly brought illumination to the darkened club room.

The crowd that gathered in the "400" Room that cold winter evening may have expected to hear "Poor Butterfly," the season's hit tune, rendered in the style of the larger, softly playing string groups popular at the time; or one of the many Hawaiian orchestras that had appeared there in Gus Edwards' revues, such as "Jonia and her Sister, the Heavenly Hawaiian Twins, and their South Seas Troubadours."

It is not difficult to see how they were momentarily stunned when, after a short introduction by the manager, the wavy-haired bandleader blew a few "licks" on his cornet, stamped his energetic foot twice, and opened up with a steaming version of "Tiger Rag," followed by his lively companions. It was as if a bomb had exploded within the room. The powerful shock waves of Tony Sbarbaro's big parade drums shook the walls; Edwards' trombone blasted and slurred with brassy bass notes that rattled every champagne glass in the room; the strident screams of Shields' clarinet echoed down the mirrored corridors and made people in the street stop to look about.

The effect on the uninitiated was summed up quite vividly, although with not too much accuracy, in a newspaper article published a few months later. The following, by F. T. Vreeland, appeared in the New York *Sun*, November 4, 1917, while the Dixieland Band was still engaged at Reisenweber's:

The young man with a face that seems to have grown florid from blowing his cornet to the point of apoplexy looks around at his handful of fellow players commandingly and begins thumping earnestly with his fashionably shod foot and instantly the whole pack is in full cry. The musical riot that breaks forth from the horns and variants of tin pan instruments resembles nothing so much as a chorus of hunting hounds on the scent, with an occasional explosion in the subway thrown in for good measure.

It is all done in correct time—there is no fault to be found in the rhythm of it. Even though the cornetist is constantly throwing in flourishes of his own and every once in awhile the trombonist gets excited about something and takes it out on the instrument, their tapping feet never miss step. The notes may blat and collide with a jar, but their pulses blend perfectly. In fact, they frequently inject beats of their own between the main thumps just to make it harder for themselves, yet they're always on time to the dot when the moment arrives for the emphatic crash of notes.

But it takes a good deal of straining of one's aesthetic sense to apply the word music to the resultant concussion even as moderns understand music after years of tympanic education by Strauss and the more flamboyant school of Russian composers. The performers have no score before them, indeed all of them are playing by ear, so their art isn't tied down to any mere earthy notes, and they go soaring. Consequently, the melody that they are ostensibly playing dies an untimely death from drowning.

Occasionally the cornetist makes valiant efforts to resuscitate it, but he is only one against several, all of them determined, so after a mad spurt, he gives up and goes careering off on a wild spree of his own. The clarinetist whoodles and whines, the trombone chokes and gargles, the violins snicker and shriek, the piano vibrates like a torpedo boat destroyer at high speed in an endeavor to make itself heard above the tumult; and the drum, labored by a drummer who is surrounded by all the most up-to-date accessories and instruments of torture, becomes the heavy artillery of the piece and makes the performance a devastating barrage . . .

The author of this imaginative article seems to have been somewhat carried away, the exact extent indicated by his account of "snickering" and "shrieking" violins. There were no violins in the Dixieland Band. However, it is an excellent example of the state of bewilderment suffered (or enjoyed) by those who had no warning of this new brand of music. The observation that "they frequently inject beats of their own between the main thumps" reveals a style of syncopation so novel that even newspaper columnists were at a loss for words. And if the melody seemed to die "an untimely death from drowning," or the cornetist occasionally went "careering off on a wild spree of his own," it was because improvisation in dance music was an art virtually unknown to New York. The rhythm, tempo, counterpoint, and volume of this band were so completely radical that immediate acceptance was impossible. People who had come to dance merely stood and gazed, first at the musicians, then at themselves. Other customers banged on their plates with silverware and yelled: "Send those farmers back to the country!"

Once more the manager stepped to the bandstand and nervously announced: "This music is for dancing."

The band played another selection, somewhat slower in tempo. A few pioneering couples gave it a try—others followed. Before this historic performance was an hour old, its audience was intoxicated with jazz.

The word spread quickly. Newspapers, in their age-old tradition, carried exaggerated stories. Friends told friends. Night-clubbers flocked to Reisenweber's to be frightened by this compact little group of rebel musicians.

From that evening on, the "400" Club Room

seldom closed before eight o'clock in the morning. One particular private party lasted until noon the next day. LaRocca and his men played seventeen hours straight at this affair, including their regular performance and the party that followed, stopping only long enough for an occasional drink of water.

Reisenweber's raised its cover charge and food prices. The band's pay was upped from $750 to $1000 per week, a miraculous salary for the period; this figure was often exceeded in tips—five, ten and fifty dollar bills that were stuffed into the "sugar can" by wealthy patrons. (The "sugar can" had been brought along from New Orleans, where it had once served to collect enough pennies and nickels to buy the members an occasional cheap meal.)

Ragtime's lusty successor had finally completed its evolution from "jass" to "jasz" to "jaz" and, in the *New York Times* of February 2, 1917, we find the first appearance of the word spelled "jazz." Reisenweber's ad on the amusement page of that issue vaunted "The First Eastern Appearance of the Famous Original Dixieland 'JAZZ BAND.'" Subsequent ads of February 9 through 13 boasted, "An Overnight Furore—The Fad of the Hour." LaRocca avers that the word "jass" was changed because children, as well as a few impish adults, could not resist the temptation to obliterate the letter "j" from their posters. By the end of February the city was full of "jazz" bands, most of them in name only. Reisenweber's felt obliged to blazon, in electric lights that could be seen across Columbus Circle: "The Original Dixieland Band—Creators of Jazz."

Reisenweber's emergence as the center of the jazz
world is revealed in *Variety* for March 19, 1917:

Music is becoming more and more potent and prominent
among the cabaret attractions. Gingery, swinging music is
what the dancers want, and it is even looked for by those who
do not dance. A group of men, the other evening, each knowing
only too well all the cabarets of New York, decided the best
restaurant orchestras in the city are Rector's, Healy's, and the
Tokio's. These orchestras get nearer to the legitimate "jazz
stuff" than any of the others. The genuine "jazz band" at Reisenweber's, however, notwithstanding the sober opinion of it,
appears to be drawing business there. Late in the morning the
jazzers go to work and the dancers hit the floor, to remain there
until they topple over, if the band keeps on playing. It leaves
no question but what they like to dance to that kind of music
and it is a "kind." If the dancers see someone they know at the
tables, it's common to hear, "Oh, boy!" as they roll their eyes
while floating past, and the "Oh, boy!" expression probably
describes the Jazz Band music better than anything else could.

Among the first musicians who came there to catch
the new style in music were Henry Busse, Ross Gorman,
Vincent Lopez, and Earl Fuller. Fuller lost no time in
converting to jazz. His "DeLuxe Orchestra" was playing
at Rector's, and the very next day after the Dixieland
Band had opened the "400" Room, he cut his personnel
down to a five-piece combination and adapted the jazz
style. According to Ernest Cutting, who was general
manager for Fuller from 1916 to 1920, "We organized the Earl Fuller Jazz Band to compete with the Original Dixieland Jazz Band which was functioning at
Reisenweber's." Earl Fuller led his band on piano, other
members including Walter Kahn (cornet), Harry Raderman (trombone), John Lucas (drums), and the up-and-
coming Ted Lewis (clarinet).

The phonograph records of "Earl Fuller's Famous Jazz Band"[1] made in early 1918—a full year after the first recording of the Original Dixieland Jazz Band—are excellent examples of the first attempts at imitating jazz. They managed to sound like the Dixieland Band, in a crude way, with their prominent counterpoint, but the rhythm was entirely different. The cornetist handled his instrument with remarkable ease and technical skill, but the style would be more fitting in a "Merchant of Venice" concert solo. Trombonist Raderman limited himself to a continuous series of long glissandi; and Ted Lewis, not yet conversant with the liquid style of Larry Shields, trilled most of the time. The drummer used a military style with frequent rolls on the snare drum and, like Sbarbaro, employed a "minstrel style" wood-block on choruses. The Dixieland Band's stock ending, the "dixieland tag," faithfully concluded every number. The general effect was that of a concert band in Central Park rendering a jazz tune especially arranged for a military combination.

One of the most sincere advocates of the new music was a little-known pianist by the name of Jimmy Durante. In 1917, more than a decade before the organization of that great vaudeville team of "Clayton, Jackson & Durante," Jimmy was holding forth at the Alamo Café, located in the basement of a burlesque house on 125th Street, where he played piano and sang songs in

[1] Victor No. 18321—"Slippery Hank"
                "Yah-de-dah"
Victor No. 18369—"The Old Grey Mare"
                "Beale Street Blues"
Victor No. 18394—"Li'l Liza Jane"
                "Coon Band Contest"

a voice that sounded like someone writing on a black-board with an ice pick. He frequently stopped in at Reisenweber's after his own show, sometimes as late as five o'clock in the morning, and once invited the whole band over to the Alamo for a late snack. The snack turned out to be a veritable banquet, with a long table set with everything from salami to *pâté de foie gras.* After the feed, the "Schnoz" sat in on piano with the band, while quiet Henry Ragas, by himself in the corner, took generous advantage of "drinks on the house." Durante, like all other ragtime pianists of the day, had trouble catching the rhythm, but he was determined to form his own jazz band. He asked about other New Orleans musicians. LaRocca suggested Frank Christian and Achille Bacquet. Durante followed this recommendation and organized a five-piece band that was billed at the Alamo intermittently for eight years, playing at various resorts during the summers. The outfit, featuring Frank Christian (cornet), Achille Bacquet (clarinet), Frank Lotak (trombone), Jimmy Durante (piano), and Johnny Stein (drums), was a particular sensation at the College Inn on Coney Island.[2]

The fame of the Original Dixieland Band had now begun to spread across the continent. An army of energetic young men in riding breeches and caps worn backwards descended upon Reisenweber's. They represented

[2] Recordings by Jimmy Durante's Original New Orleans Jazz Band:
> Gennett No. 4508—"Jada"
> > "He's Had No Lovin' For a Long, Long Time"
> Gennett No. 9045—"Why Cry Blues"
> OKeh No. 1156—"Ole Miss"
> > "Jada"

the World Film Corporation of Hollywood, California, and they moved their hand-powered cameras and battery of arc lights into the Columbus Circle restaurant to film a comedy entitled *The Good For Nothing*, starring Charlie Chaplin.[3] The Original Dixieland Band appears in a particularly animated slapstick night club scene. Movie audiences throughout the nation were, for the first time in history, to see—but not hear—LaRocca and the innovators of jazz.

Manager Max Hart was deluged with a hundred offers for the Original Dixieland Jazz Band—dances, Broadway musicals, vaudeville tours, conventions— more jobs than the band could handle in a single life-time. At a one-night benefit on the stage of the Century Theater, the Dixieland Band appeared between Enrico Caruso and Billy Sunday. Publicity continued at its peak. Sheldon Brooks, composer of "Some of These Days" and "Darktown Strutters' Ball," visited Reisen-weber's and was so thrilled that he went directly home and wrote "When You Hear That Dixieland Jazz Band Play:

Verse:
Mister Sousa has a reputation,
    But not for syncopation,
He plays all those high class marches and operas grand.
There's a jazzy band that's got me dippy,
    I'll tell you it's some band.
I'm going to tell you all about it,
    They call it Dixieland.

[3] Copyright 1917, registration LU11783 (unpublished motion picture photoplays), Library of Congress. Produced by William A. Brady, directed by Carlyle Blackwell. Title changed from "Jack the Good for Nothing."

Chorus:
Folks have you heard that Dixie Jazz Band? Say! It's a bear,
    That boy can play cornet, you bet,
    With a feeling so appealing,
They'll make you go and get your dancing shoes,
They sway and play the Livery Stable Blues.
    I love to hear that trombone moan, so beautiful.
Now if you get blue and melancholy, ask that leader to play
That jazzy every day, to drive the blues away to stay,
    Deacons, preachers, Sunday school teachers, will have to
        sway
    When they hear that Dixieland Jazz Band play.[4]

[4] Copyright 1918 by Will Rossiter, Chicago.

6

Tin Horns and Talking Machines

It is doubtful that even the imaginative Thomas A. Edison knew just what he was starting when he yelled, "Mary had a little lamb!" into a tin funnel one day in 1877 and thereupon made history by recording his voice on a rotating cylinder.

Few people, and certainly no musician of high repute, took this toy seriously until the invention of the disc record by Emile Berliner in 1896. Even then, with large-scale mass production a practicality, reputable artists were slow in adapting to the new medium. It was seven more years before opera singers could be persuaded to transcribe their voices in the scratchy wax.

The Columbia Gramophone Company was the first American concern to accomplish this cultural revolution. In 1903 they began recording the voices of such international operatic personalities as Ernestine Schumann-Heink, Suzanne Adams, Antonio Scotti, Edouard de Reszke, Giuseppe Campanari, and Charles Gilibert. The British Gramophone and Typewriter Company ("His Master's Voice") was not to be outdone. In that same year its United States affiliate, the Victor Talking Machine Company (now RCA-Victor), retaliated with an ambitious program of opera recordings including Ada Crossley, Zelie de Lussan, Louise Homer, Robert Blass, and—as the final clincher—the great Enrico

Caruso. Victor had soon stolen the thunder from Columbia.

In January, 1917, Victor was leading phonograph sales with Sousa's Band and Enrico Caruso, while Columbia was searching desperately for a means of regaining its lost favor. Then someone mentioned that latest explosion on Broadway, the Original Dixieland Jass Band. Less than a week after their spectacular opening at Reisenweber's, the New Orleans musicians were under contract to make the world's first jazz phonograph record.

From the very beginning the Columbia people did not seem to grasp the idea of jazz. Although the Dixieland Band had a repertoire of a dozen original compositions, Columbia officials did not believe that these tunes were popular enough to promote sales. After all, who had ever heard of "Tiger Rag," "Ostrich Walk," "Livery Stable Blues," or any other of these crazy animal novelties? The company therefore decided upon "Darktown Strutters' Ball" as the "selling" side of the record, and, in keeping with established practice, backed it up with an unknown tune that had been forced on them by the publishers. "Indiana" ("Back Home in Indiana") had been given little chance for success and it was hoped that "Darktown" would help it along.

The Original Dixieland Band had received the recording offer purely on the strength of their reputation at Reisenweber's. It is doubtful if Columbia, which had never even recorded ragtime and had not yet been exposed to jazz, knew exactly what it was purchasing. It is easy to imagine the state of confusion which resulted when the band started playing. The volume of this

thunderous group performing in a small studio designed for soloists and string quartets shook the very
foundations of the building. The interweaving strains of
jazz bounced from wall to wall until the resultant rever-
berations became one continuous din. The recording
director closed the door to his office from the inside.
A gang of carpenters, who were building shelves in the
studio, laughed and threw their tools about the room
to contribute to the bedlam.

After two numbers the musicians were paid their
$250 and ordered from the studio. Columbia had
washed its hands of jazz. The master record was filed
away for business reasons and forgotten. Subsequent
Columbia releases featured the Chicago Symphony Orchestra under the direction of Frederick Stock, the Cincinnati Symphony directed by Ernst Kunwald, and the
New York Philharmonic under Josef Stransky.

The southern musicians simmered for awhile
under this stinging defeat, then, with a vengeance not
concealed, went directly to Columbia's competitor, the
Victor Talking Machine Company. Here, on February
26, 1917, they recorded their two most popular compositions, "Livery Stable Blues" and "Dixieland Jass
Band One-Step" (Victor No. 18255).

From a technical standpoint Victor succeeded
where Columbia had failed. In order to understand this,
we must first consider the accoustical problems con-
fronting sound engineers in those primitive days of
mechanical recording. Before the invention of the
vacuum tube and the subsequent development of the electronic amplifier, no efficient means existed for the amplification of sound. Amplitude was the prime objective. In

order to attain sufficient power to drive the stubborn recording stylus, everything, including tonal fidelity, had to be sacrificed. Quality was but a secondary consideration. A huge tin horn was used to collect the sound and concentrate it upon the diaphragm of the recording head. This monstrous funnel of sheet metal—known as the "pickup horn"—became the central and dominant feature of every recording studio.

The accoustical problems presented by a jazz band were new to the industry. The engineers at Columbia were obviously baffled by LaRocca's unorthodox squad of music-makers. In recording opera singers, they had been able to exercise some measure of control. On particularly loud notes, for example, the singer was instructed to draw away from the pickup horn to avoid "blasting." Conversely, he was literally shoved down the mouth of the horn in order to save low notes or those lacking reproduction qualities. This process was far less dignified than an appearance on the concert stage, and some of the more famous artists, not wishing to be soiled by the greasy hands of the mechanics, actually hired assistants to help guide them back and forth before the horn.

The Original Dixieland Jazz Band, however, did not lend itself to control of this or any other sort. Because of the band's volume, the engineers feared distortion in placing the musicians too close to the pickup horn. But in placing them fifteen feet away, an echo was produced that turned the contrasting voices of the jazz band into a meaningless howl.

The problem was solved by the sound engineers at Victor, who succeeded in transcribing the band with great clarity and sharpness. They placed the musicians according to the recording strength of their instruments, and many test records were made before proper balance was attained. LaRocca was located about twenty feet from the pickup horn, while Sbarbaro wielded his drumsticks about five feet behind him. (The bass drum was not used on this record because of its tendency to "blast.") Edwards' powerful trombone was only twelve to fifteen feet from the horn, accounting for its prominence. Clarinetist Shields stood about five feet away, and Ragas, on piano—the instrument least likely to be heard—was closest of all. It is important to remember, however, that these distances were not determined by the relative volume levels of the instruments. Because the sensitivity of the recording apparatus varied widely from one tonal range to another, certain instruments were more easily picked up than others. In actuality, LaRocca, Edwards, and Shields were very evenly matched in loudness.

LaRocca describes the first recording session in these words: "First we made a test record, and then they played it back for us. This is when they started moving us around in different positions. After the first test record, four men were rushed in with ladders and started stringing wires near the ceiling. I asked them what all these wires were for, and one of the men told me it was to sop up the overtone that was coming back into the horn. The recording engineer at Victor had the patience of a saint. He played back our music until it sounded right." [1]

[1] From a note to the author, written on the back of the first-draft manuscript. Charles Souey was the patient Victor engineer.

But despite these efforts at balancing the instruments, the trombone and clarinet dominated the final
records. LaRocca attributes this to nervousness on the
part of his fellow musicians, who were inclined to play
louder on the real "take."

The stamping of a foot would be heard very
clearly, and at this time they had not yet discovered a
method of "erasing" an unwanted sound from a record.
For this reason LaRocca was not allowed to "stomp off"
his band in the usual fashion. Instead, the musicians
were instructed to watch the red signal light, count two
after it came on, and then begin playing. It is indeed
miraculous that they were able to start out together, and
even more of a wonder that they immediately fell into
the same tempo.

The world's first jazz phonograph record was boldly
launched with a special issue of the Victor Record Review, dated March 7, 1917. The company, after first
warning the customer that ". . . a Jass band is a Jass
band, and not a Victor organization gone crazy," went on
to introduce the new type of record in this vein:

Spell it Jass, Jas, Jaz, or Jazz—nothing can spoil a Jass
band. Some say the Jass band originated in Chicago. Chicago
says it comes from San Francisco—San Francisco being away
off across the continent. Anyway, a Jass band is the newest
thing in cabarets, adding greatly to the hilarity thereof . . .
Since then the Jass band has grown in size and ferocity, and
only with the greatest effort were we able to make the Original
Dixieland Jass Band stand still long enough to make a record.
That's the difficulty with a Jass band. You never know what
it's going to do next, but you can always tell what those who
hear it are going to do—they're going to "shake a leg."

The Jass Band is the very latest thing in the development

of music. It has sufficient power and penetration to inject new life into a mummy, and will keep ordinary human dancers on their feet till breakfast time . . .

The "Dixieland Jass Band One-Step," later to become known as the "Original Dixieland One-Step," was intended to be the feature attraction on this disc, although its reverse side proved to have more popular appeal. Like all the one-steps composed and rendered by this band, it is delivered at tremendous speed. Here indeed is all the martial timbre of reed, brass, and percussion. LaRocca's horn is confident, driving, yet remarkably controlled. His lip slurring in all choruses is masterful. Shield's manipulation of his clarinet through rapid, involved, polyphonic passages is breathtaking, and at a climactic point in each chorus the clarinet screams with all the excitement of a roaring crowd—it is as if someone had just scored a touchdown. In fact, the extraordinary teamwork of Edwards and Shields at the peak of this psychological build-up is well worth the listener's careful attention. Edwards is inspired to the point where he takes over the melody; or to be more specific, his tailgate *becomes* the melody. It runs rampant for about three bars until Shields scores a definite climax with his simple but well-timed glissando and Sbarbaro emphasizes the point with a dramatic cymbal crash.

Although real dixieland jazz is considered a purely abstract art, as opposed to the descriptive or programmatic art of classical music, there are intervals when jazz does approach the programmatic. The chorus of the "Dixieland Jass Band One-Step" is one of these moments. Something—some violent and thrilling spectacle—is being described by this passage. But music is different things to different people, and the listener is best left to his own interpretation.

The "B" side of this platter, the controversial "Livery Stable Blues,"[2] surprised both the band and the recording company with its overwhelming popularity. It was responsible for the sale of more than a million copies of this issue.[3] As such, it surpassed anything so far recorded by either Caruso or Sousa's Band and established a new sales record for the company. That it was instrumental in spreading jazz throughout the country cannot be denied. Even today, with the nation's population increased by more than 40 per cent and the number of phonographs multiplied possibly a dozen times, a one-million sale is considered the occasion for a special award: the gold record. And yet there are jazz histories which do not mention this record.

"Livery Stable Blues" was among the earliest of the LaRocca compositions. As previously mentioned, it resulted from an improvisation on a sacred anthem, "The Holy City," and it was developed by changing one chord and deleting two measures. In the clearly defined fugues of the verse, we are constantly reminded of the influence of the French Opera in New Orleans, where LaRocca spent many evenings of his youth as an arc light attendant. In the verse the cornet makes a statement and is answered by the clarinet, first politely and then again with a degree of impatience. The trombone enters into this "conversation" of instruments, contribu-

[2] See Chapter 7 for details.
[3] Letter to the author from E. E. Oberstein, Manager, Artists and Repertoire, RCA Victor, dated September 22, 1937.

ting a third part and—so to speak—adding its own opinions.

The animal imitations which comprise the three-bar breaks of the chorus are incredibly lifelike and may have been contributing factors in the record's popularity. Shields executes the rooster crow on clarinet, LaRocca uses his cornet to whinny like a horse, and Edwards contributes a brash donkey bray on trombone. These effects have been repeated thousands of times over by other musicians but somehow never seem to equal the character and spirit of the originals.[4]

It was not until "Livery Stable Blues" had become a smash hit that Columbia recovered the master from its dead files and made pressings of "Darktown Strutters' Ball" and "Indiana" on A2297. Following on the heels of the Victor release, it was a moderate success and 2903 was issued on English Columbia. But the Columbia recording can in no way be considered a work of art. Lacking the spirit and vitality of the first Victor disc, it is understandable why Columbia didn't release it until after its competitor had made the Dixieland Jazz Band a familiar name to all music stores. Its importance lies in the fact that it was first.

It was on this historic Columbia record that the Dixieland Band introduced "Indiana" ("Back Home in Indiana") to the phonograph audience. The band members reported to the publisher's office where they were assiduously taught the melody. After it had been impressed on their minds by a pianist who played the tune

[4] In another of the band's more popular numbers, "Walkin' the Dog," Shields used his clarinet to imitate the high-pitched "yipe" of a Pekinese. It was never recorded.

over and over, they left the publishing house and headed for Columbia's studios,[5] humming the tune en route so that it would not be forgotten before they arrived. It is interesting to note how their interpretation differs from the composer's version. LaRocca confides that he forgot a great deal of it before he reached the studio but would rather have played one of his own numbers, anyway. If nothing else, the rendition affords an excellent opportunity for the study of Tony Sbarbaro's woodblock technique, for this stands out on the record as if it were the feature attraction. The "back room" piano of Henry Ragas, almost completely obscured on all other records, is also plainly heard on several occasions.

"Darktown Strutters' Ball," on the reverse side, reveals the pains of a band in its efforts to adjust to an unfamiliar vehicle. Some parts of this tune are improvised to the point where they seem to become an entirely different melody. Adding to the group's discomfort was the handicap of playing in the wrong key. They had rehearsed the piece in the key of "C," but LaRocca, at the mercy of his peculiar musical memory, started off in "D." His colleagues had no choice other than to follow suit, for they had known the futility of fighting the determined cornetist. "I didn't realize anything was wrong," comments LaRocca, "until we had finished the number and Edwards and Shields tried to wrap their instruments around my neck!"

Meanwhile, the success of "Livery Stable Blues" continued unabated. The sales of this record in New

[5] Located on the fourteenth floor of the Manufacturer's Trust Company Building on Columbus Circle at 59th Street and Broadway, presently the site of the New York Coliseum.

Orleans reached unbelievable proportions and were largely responsible for influencing similar instrumental combinations, many of whom came north several years later. Among these was Louis Armstrong, who recollects in his autobiography *Swing That Music:*

> Only four years before I learned to play the trumpet . . . the first great jazz orchestra was formed in New Orleans by a cornet player named Dominic James LaRocca. They called him "Nick" LaRocca. His orchestra had only five pieces, but they were the hottest five pieces that had ever been known before. LaRocca named this band, "The Old Dixieland Jazz Band." He had an instrumentation different from anything before—an instrumentation that made the old songs sound new . . . They all came to be famous players and the Dixieland Band has gone down now in musical history. Some of the great records they made, which carried the new jazz music all over the world in those days were: "Tiger Rag" . . . "Ostrich Walk" . . . "Livery Stable Blues" . . . LaRocca retired a few years ago to his home in New Orleans but his fame as one of the great pioneers of syncopated music will last a long, long time, as long, I think, as American music lives.[6]

And how did New Orleans feel about the revolutionary new music that had been created by five of its local boys? If there is any doubt that jazz was strange and unheard of in the Crescent City, even as late as two years after the Original Dixieland Jazz Band went north, the following editorial in the New Orleans *Times-Picayune* of June 20, 1918, should dispel the illusion:

JASS AND JASSISM

Why is the jass music, and therefore the jass band? As well ask why is the dime novel, or the grease-dripping doughnut? All are manifestations of a low streak in man's tastes that

[6] *Swing That Music* (New York, 1936), pp. 9-10.

has not yet come out in civilization's wash. Indeed, one might go further, and say that Jass music is the indecent story syncopated and counterpointed. Like the improper anecdote, also, in its youth, it was listened to blushingly behind closed doors and drawn curtains, but, like all vice, it grew bolder until it dared decent surroundings, and there was tolerated because of its oddity . . .

In the matter of jass, New Orleans is particularly interested, since it has been widely suggested that this particular form of musical vice had its birth in this city, that it came, in fact, from doubtful surroundings in our slums. We do not recognize the honor of parenthood, but with such a story in circulation, it behooves us to be last to accept the atrocity in polite society, and where it has crept in we should make it a point of civic honor to suppress it. Its musical value is nil, and its possibilities of harm are great.

A few days later, the *Times-Picayune* received and published a letter from an angry reader who resented the editorial remarks about his favorite delicacy—the grease-dripping doughnut!

7

The Strange Case of the "Livery Stable Blues"

Besides being the world's first jazz phonograph record, "Livery Stable Blues"/"Dixieland Jass Band One-Step" was probably also the first record in history to instigate a lawsuit for each of its sides. If laid end to end, the legal documents resulting from the sale of this record would extend the full length of Canal Street and would probably tie up traffic for two weeks.

The Dixieland Jazz Band that ventured northward from the Sunny Southland in 1916 was composed of five innocents who had never heard of a copyright and never suspected that anyone would be so mean as to steal another man's tune. They were unfamiliar with the ways of the North, but in the words of LaRocca, "We sure learned fast!"

That "Livery Stable Blues" was the rightful property of the Original Dixieland Jazz Band, both in name and substance, cannot be doubted. The tune was composed by LaRocca in 1912 and was brought to Chicago with Stein's band in March of 1916 (see Chapter 3). It carried no name until it was introduced by this band a few weeks later, at which time Ernie Erdman, the house pianist at Schiller's, contributed the title.

But the Victor Talking Machine Company found "Livery Stable Blues" an objectionable title. They could not, they felt, allow such a vulgarity to appear along-side time-honored and respected operatic titles on the pages of the Victor record catalogue. So J. S. Mac-Donald of Victor suggested "Barnyard Blues" as a polite substitute. On April 9 Max Hart, the band's agent, copyrighted the composition under this title (and in his own name!).[1]

Everything would have turned out fine if someone somewhere along the line had not made one of those clerical errors, one of those boners that turns up daily in every business and often jams the clockwork of our complex machinery of civilization. Due to somebody's slip-up, the record was labeled and issued as "Livery Stable Blues," therefore legally unprotected and free for the world to copy.

Alcide "Yellow" Nunez, former clarinetist of the Original Dixieland Band and at this time playing with Bert Kelly's band at Chicago's Casino Gardens, stumbled into this golden opportunity. He had heard the record, but when he tried to purchase arrangements he discovered that the composition had never even been reduced to writing! Checking with the Library of Congress, he learned that no such title as "Livery Stable Blues" had ever been registered.

Yellow rubbed his hands together in anticipation of a real killing. There was nothing dishonest, he ration-

[1] Registration No. E-404182 (unpublished composition), Library of Congress. Surprisingly enough, this copyright was not secured until a month after the record was placed on sale. Apparently no one considered this of importance until advance sales indicated a demand for sheet music and band arrangements. Then the red tape of writing up a piano arrangement of the piece and submitting it for copyright added to the delay.

alized, in publishing a song he used to play at Schiller's with Stein's Dixie Jass Band. Nick and the boys wouldn't care—they obviously weren't interested in publishing it, anyway. And with the Original Dixieland Band presently providing free advertising of the tune from coast to coast and plugging it nightly at Reisenweber's, he could make himself a fortune. Besides, didn't Nick throw him out of the band at Casino Gardens? This would be nice, juicy revenge. Consequently, Nunez contacted Roger Graham, a Chicago music publisher, and in 1917 "Livery Stable Blues" was first printed and circulated by that company. On the published copies, Alcide Nunez and Ray Lopez (Tom Brown's former cornet player, then playing with Bert Kelly's band) were given as composers and credited with being members of the Original Dixieland Jazz Band. The cover carried a reference to the Dixieland Band's Victor record.

LaRocca went into action immediately upon learning of the theft and publication of his brainchild. A well-known New York theatrical attorney, Nathan Burkan, was engaged and, following his advice, the members of the Original Dixieland Jazz Band filed an injunction to enjoin Roger Graham from continuing publication. Due to the customary delays, the case was not called for another five months. In the meantime, New York publisher Leo Feist brought out the number as "Barnyard Blues," listing D. J. LaRocca as composer, and carrying the warning "Dealers Are Subject to Damages for Selling or Having Copies of the Spurious Edition In Their Stock." But the discrepancy between the title of the record. and the title of the sheet music was a severe handicap to the sales of the latter. So the musicians brought suit against the Victor Talking Machine Company, claiming $10,000 loss in sheet music royalties because of the mix-up. In July LaRocca took his band to the Aeolian Company for a series of four records (discussed in the next chapter). The Victor case was settled out of court for $2,500 plus the band's agreement to return to the fold and record exclusively for Victor.

On October 2, 1917, the case of LaRocca, Edwards, Shields, Ragas, Sbarbaro and Hart versus Roger Graham was heard in the United States Federal Court, Northern District of Illinois. Traditionally on the alert for anything that could be turned into slapstick comedy were the local newspaper reporters, who stampeded into the Chicago courtroom with moistened pencils and over-stimulated imaginations. From the beginning, jazz was a naughty word in that city and deserving of appropriate treatment. You could no longer ignore jazz musicians, but you could still make fun of them.

The Chicago *Daily News* of October 11, 1917, in an article headed "DISCOVERER OF JAZZ ELUCIDATES IN COURT," sarcastically portrayed LaRocca as "the Jazz Kid himself, the giddy boy whose brain first got the big idea" and reported that he "bounced" into the courtroom "all rigged out in a pair of cloth-topped patent leathers, a purple striped shirt and a green tunic." Then, according to the writer, "he identified himself as the genuine Columbus of the Jazz, the Sir Isaac Newton of the latest dance craze." LaRocca's southern vocabulary became strictly Lower East Side for the purposes of this colorful article, with as much of a Jersey accent as could be suggested through the medium of newsprint.

The headlines hawked:

JAZZ BAND MAY PLAY IN COURT (Chicago *American*)

JAZZ BAND WILL WAIL 'BLUES' IN COURTROOM (*American Inquirer*)

BARNYARD SYNCOPATION TO EDIFY JUDGE (*American Inquirer*)

'JAZZY BLUES' TO MOAN LURE IN U.S. COURT (Chicago *American*)

But Federal Judge George A. Carpenter changed his mind about letting a jazz band play in his courtroom. Things were embarrassing enough, he thought, without having donkey cries and horse whinnies echoing up and down the hallway of the Federal Building. Daily he struggled to retain his professional dignity as the scene in which he had unwittingly and helplessly become a principal actor threatened to resemble a comedy skit on a burlesque circuit.

From the very beginning the case was doomed to end in total confusion, for the actors in this little play could never meet on common ground. The stage was filled with musicians who could not explain what they composed or played because they could not read music; highly educated music authorities who could not understand the musicians; lawyers who could not understand the authorities; and a judge who was utterly disgusted with the whole business.

Prosecuting attorney Burkan called in experts to prove that "Livery Stable Blues" and "Barnyard Blues" were one and the same, and that this composition had originated in the Original Dixeland Jazz Band, a cooperative organization claiming joint ownership.

Theodore F. Morse, composer of nearly five hundred popular songs, including "Blue Bell" (one and a half million copies sold) and "Mother" ("Put Them All Together They Spell . . ."), testified that the disputed blues compositions were similar in all respects: "The introduction is identical as to harmony, rhythm and only slightly changed in melody in the third bar. The first strain is practically identical and also the second and third strain. I think these two numbers could be played at the same time and no difference would be noted that they were different compositions."

Lee Orean Smith, a composer and music editor employed by Leo Feist and Company, took a similar view: "The theoretical structure is practically the same, both as to melody, not so much identical as to harmony, but the rhythmical feeling is practically identical, therefore the metre is conceived in the same strain . . . the tone progression is practically identical, that being what constitutes the melody. There are slight variations which to the layman's ear would be imperceptible . . . They could be played together with almost complete satisfaction to the ear, as regards melody. Discords would occur at points where harmony varies."

Graham and Nunez based their defense on the premise that all blues, having the same chord progression, were alike, and therefore plagiarism was impossible. They argued still further that "Barnyard Blues" was copied from an older tune called "More Power Blues."

Miss May Hill, a music critic, was called to testify for the defense. She flaunted an array of sheet music before the judge, including such examples as "Chicago

Blues," "Alabama Blues," "New Orleans Blues," and "Livery Stable Blues," contending with a great show of technical palavering that they were all alike.

"Could they all be played at once?" inquired his honor.

"They could, and produce perfect harmony," Miss Hill asserted. The judge again ruled out the suggestion that it be given a try in the courtroom.

But the star of the program appeared on the stand the next day. All sorts of technical dissertations had been listened to by the judge, but none rocked the court-room so much as the testimony of Alcide Nunez: "Jedge, blues is blues!"

The reporters licked their pencil points. Here was something worth writing about. As Nunez unwound his invective, they were never less than two words behind him—and sometimes several sentences ahead. The Chicago *Daily News* for October 12, 1917, in a news item headed "NOBODY WROTE THOSE LIVERY STABLE BLUES," described Alcide's testimony as follows:

"You see," he said, "nobody wrote the 'Livery Stable Blues.' Naw. Nobody writes any of that stuff. I invented the pony cry in the Blues, and LaRocca, he puts in the horse neigh. We was in the Schiller cafe, rehearsin', see? And I suggests that we take the 'More Power Blues' and hash 'em up a bit. My friend, Ray Lopez, he wrote the 'More Power Blues.' All blues is alike . . .

"That's what," continued he. "We hashed up the 'More Power Blues' and put in the pony cry and the mule cry and the horse neigh, see? Then we rehearsed it for ten days, steaming it up and getting it brown and snappy. Then we had the piece all fixed.

"No, I don't read music. I'm a born musician. Yes, sir. I plays by ear exclusively. I've played in all the swell places

with Kelly's band—in the Sherman Hotel and all over. I'm entitled to the authorship of the 'Livery Stable Blues,' me and
Lopez, as much as LaRocca, that's why I went to Roger Graham and had him publish it. LaRocca done me dirt, so I says
to myself, 'He's done me dirt and I'll let him out.' He goes and
has our 'Livery Stable Blues' put on a phonograph record as
his'n. Well ain't that dirt?"

Nick LaRocca was the only member of the Dixieland Band to appear in person at the trial. The others
remained in New York and testified *in absentia*. There
is ample evidence that LaRocca recognized the publicity
values latent in the case. Throughout the many newspaper accounts of these hearings there runs an overtone
of subtle showmanship, witness the following item from
the Chicago *Journal*, October 11, 1917:

JAZZ BAND MASTERPIECE AUTHORSHIP IN DISPUTE
Dominic Claims Alcide Purloined His Tone Picture
of Emotions of Lovesick Colt
    Dominic LaRocca was once an electrician. The lure of
the cornet tore him away from the business of giving light to
the citizens of New Orleans. He roamed the streets, putting
heart and soul into his wonderful horn. Dominic realized what
spiritual messages might mean. He interpreted, through his
cornet, the sweet braying of the lonesome donkey, and the
gentle neighing of the love-sick colt.
    Such was the birth of the "Livery Stable Blues," swears
Dominic, and before Judge Carpenter in the Federal Court
today, he asked an injunction against the envious, jealous,
Alcide Nunez, clarionet [sic] player of the Original Dixie Jazz
Band. Alcide is charged with stealing the song, selling it to the
Roger Graham Music Publishing Company, and living proudly
on the income and reputation it has won him.
    It matters not that the new title of the song is "Barnyard
Blues." According to Attorney Bryan Y. Craig, his client,
Dominic, will speedily demonstrate to the court that the mellif-

luous harmonies could never have proceeded from a barnyard, but only from a livery stable.

When the thirty-five witnesses have completed their testimony, and Judge Carpenter has heard from jazz experts, cabaret owners, and livery stable keepers, the composer of the world's greatest jazz music will be identified. The "Livery Stable Blues," along with the Shakespeare-Bacon controversy, will have become history.

Ernie Erdman, Schiller's pianist, could have settled the whole matter, for it was he who originally suggested the title. But Erdman wasn't in business for his health, either. In a letter to LaRocca he explained the conditions under which he would appear as witness for the prosecution:

". . . You readily admit in your letter, as far as the "Livery Stable Blues" is concerned, that this was my title, and also that there was some transaction relative to counting me in on the number, but up to the present time, this is the first line I have ever had from you on the subject one way or another . . . Consequently you will understand my position in the matter thoroughly, and also realize how inconsistent it would be for me to do anything unless I am given my one-sixth of the royalties . . ."

Thus still another staked his claim to joint ownership and a cut on the fabulous profits of "Livery Stable Blues." But the band stubbornly refused to pay for Erdman's testimony.

The hearings dragged on for ten days, during the course of which Judge Carpenter showed visible effects. The Chicago *American* reported that he repeatedly "fled to his chambers for ice water." He confided to his friends that he was beginning to wake up in the middle of the night and hear donkey cries emanating from his butler's pantry and horse whinnies from his icebox room. He said he was determined to clear this business from the courtroom before he got roosters in his wine cellar.

On October 12, Judge Carpenter rendered his decision:

Gentlemen, there is not any law to this case otherwise than would be ordinarily submitted to the determination of a jury. It seems to me it is a disputed question of fact, and it is passed on to the Court just as any like case would be passed on to a jury.

There is a dispute between the plaintiff and the defendant, two publishers, each claiming a right to the monopoly of this song, this musical production.

No claim is made by either side for the barnyard calls that are interpolated in the music, no claim is made for the harmony. The only claim appears to be for the melody.

Now, as a matter of fact, the only value of this so-called musical production apparently lies in the interpolated animal and bird calls—that is perfectly apparent from the evidence given by all the witnesses, and in a great many unimportant features the Court has great difficulty in believing what some of the witnesses on both sides of this case have told the Court under oath, but that does not go to the real merit of this controversy.

The cat calls and animal calls were not claimed in the bill and they were not included in the copyright, so we are to exclude them in this question.

The only question is, has there been a conceived idea of the melody that runs through this so-called "Livery Stable Blues" . . .

The last witness says this "More Power Blues" is fifteen years old and the plaintiff's best witness Mr. LaRocca says it is ten years old, and from the evidence here which is in dispute —of the two young ladies who testified that these are alike and unlike, the Court cannot attach any great value to it in this case because they do not agree.

But the Court is satisfied, from having looked over the

manuscripts, that there is a very decided resemblance between the aria—the melody of "More Power Blues" and the "Livery Stable Blues."

The finding of the Court is therefore that neither Mr. LaRocca and his associates nor Mr. Nunez and his associates conceived the idea of this melody. They were a strolling band of players and like—take the Hungarian orchestras, if you will, but with no technical musical education, having a natural musical ear—quick ear and above all a retentive ear, and no human being could determine where that aria came from that they now claim was produced at the Schiller Cafe for the first time —whatever was produced by the rhythm was the result that pleased the patrons at the place, and it was the variation of the original music that accomplished the result and not the original music itself, and I venture to say that no living human being could listen to that result on the phonograph and discover anything musical in it, although there is a wonderful rhythm, something which will carry you along especially if you are young and a dancer. They are very interesting imitations, but from a musical stand-point it is even outclassed by our modern French dissonance.

And the finding of the Court will be that neither the plaintiff nor the defendant is entitled to a copyright, and the bill and the answer will both be dismissed for want of equity.

Having thus dismissed "Livery Stable Blues" as an unmusical noise and jazz musicians in general as wandering Hungarians, the judge casually slammed the covers shut on another volume of musical history.

Nathan Burkan was appalled by the decision. He assured his clients of certain victory if the case were taken to a higher court. But LaRocca demurred, indicating that the suit had already cost the band more than two thousand dollars. Burkan, however, had one more suggestion: two could play at this game. If the court found that all blues were alike and that one could be freely copied from another, there was nothing to stop the Original Dixieland Band from rehashing "More Power Blues." Anyone then proving plagiarism would automatically reverse the court's decision. In order to bring about just such a test case, the Original Dixieland Band lifted "More Power Blues," making a few minor changes and claiming it as their own number. "Mournin' Blues" (Victor 18513) was the result. Strangely enough, no one ever challenged the theft, and today "Mournin' Blues" is accepted as a standard part of the Original Dixieland Band's repertoire of original compositions.

No sooner had the "Livery Stable Blues" litigation run its unsuccessful course, than someone took a closer look at the reverse side of the platter. Song publishers Joseph W. Stern and Company (now Edward B. Marks Music Corporation) of New York claimed that the trio of "Dixieland Jass Band One-Step" had been pilfered from one of their numbers, "That Teasin' Rag," written by Joe Jordan in 1909. Jordan had since sold his copyright to Stern and so Stern brought action against the Original Dixieland Jazz Band, based on the rather weak coincidence that two bars of the trio were similar. Victor immediately withdrew all available copies of the record from its distributors and reissued them with new labels on which the phrase "Introducing 'That Teasin' Rag'" was added to the title, thereby turning the number into a medley and accounting for the discrepancy in labels that has intrigued record collectors for more than forty years.

In the meantime LaRocca beat a weary path to the door of the publishers and, in a conference that lasted four hours, succeeded in salvaging one third of the royalty rights to the number he had composed. The other two-thirds were relinquished to the company, who still publish the "Original Dixieland One-Step," although "That Teasin' Rag" has long since been discontinued.

Although "Livery Stable Blues" (alias "Barnyard Blues") disappeared from the jazz scene many decades ago, its companion "Original Dixieland One-Step" has since become the national anthem of dixieland musicians. Probably no dixieland band presently in existence cannot render a version of it on request, and as a final tribute to this great grand-daddy of jazz band compositions, the Edward B. Marks Music Corporation has recently published a folio of band arrangements known as "The Original Dixieland Concerto"—a fascinating medley of "Original Dixieland One-Step," "Jazz Me Blues," and "Ballin' the Jack"—aimed at the vast market of high school bands and orchestras throughout the country.

8

Hot Licks and Cold Wax

The Bolshevik revolution and the subsequent collapse of Russia in the spring of 1917 brought the morale of the Western nations to a new low. Effective allied propaganda had whipped the passive indignation of Americans into a fervid crusade spirit, and when Germany finally extended its unrestricted submarine warfare to the sinking of Belgian relief vessels and American merchant ships in March of that year, it was the proverbial last straw. On April 16, 1917, the United States declared war on Germany.

In New York City, the mass release of human emotions manifested itself in a number of ways. "War jitters" brought a new boom in the entertainment business, as thousands sought escape from international tensions in the fast-moving night life of Manhattan. And it was jazz music that struck the new tempo.

The Original Dixieland Jazz Band, encouraged by the immediate success of "Livery Stable Blues," moved into its greatest creative period, out of which emerged such classic jazz compositions as "Clarinet Marmalade," "Bluin' the Blues," "Fidgety Feet," "Lazy Daddy," and "At the Jazz Band Ball." These numbers, which have become "standards" with every present-day dixieland jazz band, were born at Reisenweber's during a series of jam sessions held by the band during the early hours of the afternoon. These sessions are enough to arouse the imagination of the most casual jazz fan, and what this writer would give to have been a bus-boy or waiter at Reisenweber's in the spring of 1917!

It was decided from the beginning, when the Original Dixieland Jazz Band first came into existence, that it would be a co-operative organization with all members sharing alike. For this reason royalties have always been split five ways, and even the credit of authorship distributed so that each musician's name appeared on at least one composition, whether he had contributed to it or not. This generous, brotherly philosophy eventually led to the internecine warfare covered at length in subsequent chapters.

There is much evidence that the seeds of all the compositions took root in the fertile musical mind of D. J. LaRocca. Not only do certain recurrent phrases appear in many of the numbers, indicating the ideas of one man, but the mere fact that LaRocca was the only member to compose before and after, as well as during, the life of the Dixieland Band would seem to cast doubts upon the musical prolificacy of the other members. The creative genius of LaRocca existed before the organization of the band and still continues.

It is only natural that the musician who carried the melody would have the greatest say-so in the determination of its course. The cornetist was, after all, the only member directly concerned with the strain of continuity, the musical line which is regarded as the "story" itself. The others could do nothing more than support this strain. In the sense of popular music, LaRocca is the author of all these compositions because he conceived the melody. But in the sense of classical music, especially symphony, where the theme is frequently subordinate to the overall score, the contributions of the other musicians must not be overlooked.

Certain passages exist in many of the LaRocca compositions in which the countermelody, more often the trombone part, becomes dominant. The chorus of "Fidgety Feet," for example, stripped of its counterpoint, would sound more like a hymn than a jazz number. In many cases, such as "Livery Stable Blues" and "Ostrich Walk," LaRocca instructed his musicians in just how this counterpoint was to run. The musicians themselves have all testified to this in a court of law (see Chapter 7). The whole arrangement, like the score of a symphony, developed in the mind of the composer before the first sounds were uttered.

Just where the autocracy of LaRocca ended and the group effort began is a difficult matter for conjecture. The world will never know, for there is little agreement among the surviving members on this point. But although the evidence is mainly circumstantial, it is very heavily weighted in favor of LaRocca.

Larry Shields comes to the foreground as the second most important element in the group's composing effort. His great virtuosity, coupled with the very nature of his instrument, made him the "showpiece" of the band. Most of the "breaks" in the dixieland masterpieces are therefore clarinet "breaks," designed to show off the great skill and imagination of this musician. The ideas in the "breaks" are always Larry's, and sometimes his hand is demonstrated in the body of the chorus, as in "Clarinet Marmalade." For this reason, the names La-

Rocca and Shields often appear together beneath the titles on sheet music, arrangements, and phonograph labels. (In all other LaRocca compositions, the composer is presently designated as "The Original Dixieland Jazz Band," the result of disputes that were never settled.) The relationship of LaRocca and Shields was collaboration in its highest form, and the rehearsals at which these two men worked out the details of such scintillating numbers as "At the Jazz Band Ball" must have been memorable occasions.

The highly talented but intractable Edwards, according to LaRocca's complaint, slept late in the day and rarely attended jam sessions or rehearsals. This trait perhaps suggested the title "Lazy Daddy," "Daddy" Edwards being so named because he was the first band member to become a father. Eddie had married a show-girl in Gus Edwards' review at Reisenweber's in 1917, and when little Branford Edwards was baptized, Nick appeared in the role of godfather.

The absence of "Daddy" Edwards at rehearsals was a constant source of irritation to LaRocca, who was anxious to develop a repertoire so extensive that the band could play all evening without resorting to the works of other composers. It was also found that the band was at its best in rendering its own compositions, that only the musicians themselves could provide the custom-built framework so critical to their needs. But Edwards was always confident of his ability to pick up a tune on the job, relying on his quick ear and acute memory to carry him through a new number. It is conceivable that he considered himself a superior musician, being the only member of the group who could read music. In fact, LaRocca relates how Edwards would demonstrate new numbers received from the publishers, playing them over on his trombone directly from the lead sheets until LaRocca had learned the melody.

The fact that Daddy Edwards could read music was kept a closely guarded secret for many years. For publicity purposes, the Original Dixieland Jazz Band was proclaimed far and wide as a collection of supremely talented musicians, who, in the New Orleans tradition, could not tell a dotted eighth from a hole in the ground. It is understandable, then, how a competent, formally trained reader such as Edwards might well be considered a dangerous reactionary and a threat to the sublime musical virginity of the organization. He might even have been falsely accused of playing from memory instead of by ear. That his versatility was intentionally underplayed must certainly have galled Edwards, who was known in private to mock the audience's adulation. "Oh, aren't we wonderful?" he would exclaim. "We can't read music!"

In reality, the ability to read music is neither a help nor a hindrance to the jazz musician. Whether style is influenced by the manner in which a "faker" conceives his music—the abstract patterns which take the place of musical notes in his mind's eye—is an open question for pioneering psychologists.

The Dixieland Band was set up as a co-operative enterprise, but no organization can progress without centralized leadership. By mutual agreement LaRocca had originally been appointed musical leader and Edwards the business manager. These two senior members had always considered themselves a partnership. Ed-

STEIN's DIXIE JASS BAND at Schiller's Café in Chicago, March 1916. The nucleus of the Original Dixieland Jazz Band included (left to right) Alcide Nunez, Eddie Edwards, Henry Ragas, Nick LaRocca, and Johnny Stein. Dusters served as uniforms.

THE ORIGINAL DIXIELAND JAZZ BAND in New York, January 1917. Reisenweber's demanded tuxedos. Left to right, Larry Shields, Eddie Edwards, Tony Sbarbaro, Nick LaRocca, and Henry Ragas.

THE ORIGINAL DIXIELAND JAZZ BAND at Reisenweber's. Trombonist Edwards did not ordinarily stand inside the piano.

JIMMY DURANTE's Orginal New Orleans Jazz Band (1917). Jimmy "imported" four jazzmen from New Orleans. Left to right, Johnny Stein, Frank Christian, Frank Lotak, Durante, and Achille Bacquet.

FRONT COVER of Victor catalog for March 17, 1917, advertising the world's first jazz phonograph record.

On Stage at the Palladium, unposed

The Dixieland Band at the Palais de Danse, London, 1919. The British thought it was "a musical joke hardly worth attempting." Left to right, Billy Jones, Larry Shields, Nick La-Rocca, Emile Christian, and Tony Sbarbaro.

Soothing The Savage Beasts at Central Park Zoo, April 1921.

THE DIXIELAND BAND in 1922. Robinson got mad, Shields got homesick. Left to right, Henry Vanicelli, Artie Seaberg, La-Rocca, Edwards, and Sbarbaro.

THE DIXIELAND BAND in 1936. They could still cut it. Left to right, Edwards, Sbarbaro, LaRocca, Shields, Robinson.

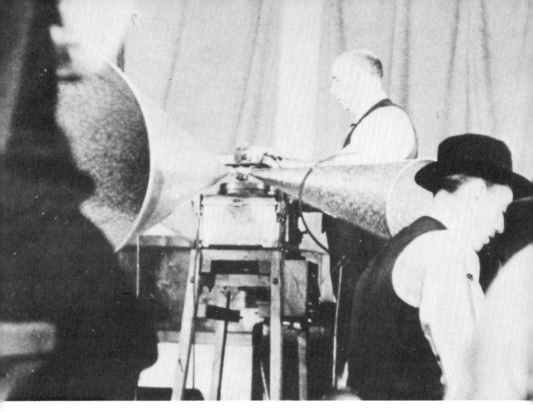

EARLY RECORDING TECHNIQUES. From the *March of Time* film featuring the Original Dixieland Jazz Band.

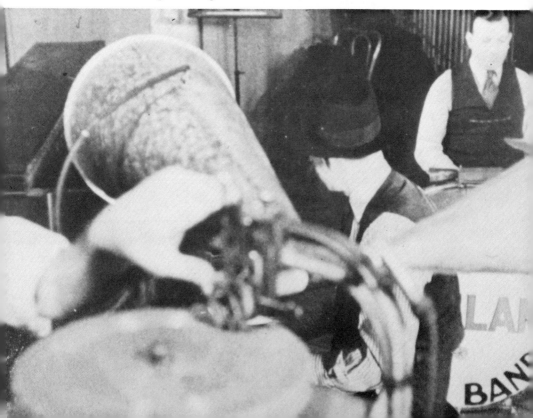

wards, with less of a southern accent, did the announcing and general talking for the band, while his partner devoted himself to the musical details. But, according to LaRocca, Edwards took the business end of it less seriously than he should have and was glad to relinquish his responsibilities. It is uncertain just when Nick LaRocca took over full leadership of the band—including the management of business affairs—but contracts and other documents pertaining to phonograph recording and music publishing, dating back to the latter part of 1917, are all addressed in his name. In addition, it was LaRocca who hired the lawyers, paid the bills, and went to Chicago to represent the band in the "Livery Stable Blues" case.

Too much importance cannot be attached to the role of LaRocca both as composer and bandleader. The very duality of this role was responsible for jazz compositions which are much more than melodies and which require more than one instrumental voice for their presentation. LaRocca was delicately sensitive to the musicians who played with him, always alert to their spontaneous improvisations, which he organized and helped develop into the counterpointed music now known as dixieland jazz. His very attitude toward music was bound to be productive of a new form.

But the germs of the ideas that led to the great 1917 compositions came into being long before the Reisenweber jam sessions. Military bands in New Orleans were the primary influence on young LaRocca, and marches determined the form and flavor of the jazz numbers that were to follow. Of paramount significance is that a peculiar quirk in his musical mind made it impossible for him to repeat a tune exactly as he heard it. His imagination was so violently active that there was no room for exact interpretation. The tunes that made such an impression on his mind at concerts in New Orleans were not remembered in the ordinary manner. When he returned home to try them out on his cornet, entirely new melodies resulted. It was not loss of memory but a continuous process of improvisation which went on in his brain.

He explains that the first part of "Fidgety Feet" came out of an attempt at playing "Georgia Camp Town Meeting," written by Kerry Mills in 1897. By examining these two compositions, it will be noted that both are of the same length and have a common chord progression. Yet they are so different in melody and rhythm that probably no one has noticed this basic similarity. "At the Jazz Band Ball" is another example of La-Rocca's method of composition in which the chorus is actually an improvisation of "Shine On Harvest Moon." Here again, the chord progressions are identical.

Nowhere was the faculty of deriving new melodies from old ones more skilfully exercised than in La-Rocca's "Tiger Rag." But here the eclectic process involved not just one established tune but several. The first two bars of the verse were a note-for-note copy of the "get over dirty" phrase so popular with New Orleans musicians of that day.[1] The second part of the verse was

---

[1] "Get over dirty" was not a full song, but merely a two-bar "lick"—something like the "shave-and-a-haircut" and "over-the-fence-and-out" routines—and had a certain esoteric significance known only to the musicians, in the manner of a private joke.

actually a simplified version of "London Bridge Is Falling Down" in stop-time. The trick was not new and was commonly used by vaudeville orchestras to accompany acrobatic acts. LaRocca's treatment, as was his custom, was slightly skeletonized.

The chord progression upon which the "Tiger Rag" choruses were built was borrowed from the "National Emblem March," but the melody has other origins. The "riffs"[2] which comprise the second, or "hone-ya-da," chorus began as a humorous imitation of the alto part in a German band. The third, or "hold that tiger," chorus evolved gradually from the "hone-ya-da" sequence through several stages of improvisation. The first and last choruses were still further improvisations.

If such odd phonetic phrases as "hone-ya-da" sound contrived and unwarranted to the reader, they were nevertheless of vital importance to the members of the Dixieland Band, most of whom could not read music and had no other way of identifying the parts of their compositions. "Hold that tiger" served the same function but did not come into use until the number was named by the recording company. In passing, it is interesting to note that the tiger roar in the third chorus was originally executed on cornet.

The genius that manifested itself in such unforgettable music obviously did not extend to the printed word. Some of the names suggested by LaRocca for his compositions would have been more suitable for race horses. Although "War Cloud" and "Belgian Doll" were

---

[2] "Riff": A simple phrase, usually of two measures, repeated for the duration of a chorus; nowadays more frequently used as the background to a solo.

timely, considering America's entry into the war and her popular sympathy for Belgium, the Victor Company, according to official correspondence, did not consider them "particularly appropriate." J. S. MacDonald of that company, who had already created utter chaos by substituting "Barnyard" for "Livery Stable," was ready with bright new titles. At his suggestion "War Cloud" became "Fidgety Feet," and "Belgian Doll" turned into " 'Lasses Candy."

The creative output of Nick LaRocca is partly revealed in Table 4. This list, representing nineteen jazz compositions, does not include his efforts in the field of popular songwriting, which resulted in "The Army Mule" (1917), "When You Roll Those Dreamy Eyes" (1917), "King Tut Strut" (1921), and "Some Rainy Nights" (1923). In addition there were any number of melodies handed out to well-known but unscrupulous lyricists of the day, only to be stolen outright or plagiarized by competitors. Such productivity may seem even more remarkable when it is considered that the composer, minus any faculty for reading or writing music, carried all these numbers around in his head until someone showed an interest in transcribing them on paper.

With so many musical ideas straining to break loose from the brain in which they were imprisoned, a certain amount of confusion was inevitable. Certain pet phrases of LaRocca's, common to several of his compositions, proved to be perilous traps for himself and the band during performances. He would sometimes start out playing one number and end up playing another, much to the consternation of his fellow musicians.

TABLE 4

Compositions of the Original Dixieland Jazz Band

DATE*	TITLE	COMPOSER(S)	PUBLISHER
1912	Livery Stable Blues	LaRocca	Leo Feist & Co.
1912	Tiger Rag	LaRocca	Leo Feist & Co.
1912	Sensation Rag	LaRocca	Leo Feist & Co.
1912	Orig. Dixieland One-Step	LaRocca	Edward B. Marks Co.
1914	Ostrich Walk	LaRocca	Leo Feist & Co.
1917	Skeleton Jangle	LaRocca	Leo Feist & Co.
1917	Reisenweber Rag	LaRocca	Unpublished
1917	At the Jazz Band Ball	LaRocca and Shields	Leo Feist & Co.
1917	Clarinet Marmalade	LaRocca and Shields	Leo Feist & Co.
1917	Fidgety Feet	LaRocca and Shields	Leo Feist & Co.
1917	Lazy Daddy	LaRocca and Shields	Leo Feist & Co.
1917	Look at 'em Doin' It	LaRocca and Shields	Leo Feist & Co.
1917	Mournin' Blues	Original Dixieland Jazz Band	Leo Feist & Co.
1918	Bluin' the Blues	Original Dixieland Jazz Band	Leo Feist & Co.
1918	Satanic Blues	LaRocca, Shields, and Christian	Leo Feist & Co.
1918	'Lasses Candy	LaRocca	Leo Feist & Co.
1920	Toddlin' Blues	LaRocca	Leo Feist & Co.
1920	Ramblin' Blues	LaRocca and Shields	Morris Music Co.
1954	Basin Street Stomp	LaRocca and Franks	Edward B. Marks Co.

* date composed

Dancers close enough to the bandstand might see Edwards drop his horn and shout to LaRocca, "Hey, Joe! Where you goin'?"[3]

During the suit against the Victor Talking Machine Company, following the "Livery Stable Blues" mix-up, the band temporarily broke off relations with the company. Max Hart secured a contract with the Aeolian Company, and in July of 1917 the band went over to Aeolian Hall to record several of their compositions (see Table 5), including the one that had already incited a federal court injunction.

Although disc records had by this time replaced the older cylinders, the controversy over lateral versus vertical ("hill and dale") recording still raged—a commercial conflict somewhat comparable to the "battle of the turntable speeds" in 1949, when Columbia's 33⅓-rpm long-playing record emerged victorious over RCA's 45-rpm "doughnut." In 1917 Aeolian, like Edison, was producing records with "hill and dale" grooves, in which the needle vibrated up and down instead of sideways. Both Victor and Columbia had gone to lateral recording because of the increased volume obtainable. The final result of this "war of the needles" was that lateral recording became standard, and Aeolian, with a fortune tied up in specially built phonographs, went out of business. (They reorganized later using lateral recording methods.) The old Aeolian-Vocalion records

---

[3] LaRocca was nicknamed "Joe Blade" because of his shrewd business manipulations. The phrase "Old Joe Blade, Sharp as a Razor" was coined by Eddie Edwards early in the band's career and later became the title of a popular song composed by LaRocca and J. Russel Robinson.

have since become collectors' items and never fail to intrigue listeners, partly because they can be played on a modern phonograph without producing a single sound! It is possible, however, to hear these records by turning off the phonograph amplifier and keeping an ear close to the pickup arm. Although modern phonograph cartridges do not respond to vertical motions of the stylus, a certain amount of vertical vibration is picked up and reproduced mechanically by the pickup head.

On July 29, 1917, the musicians gathered for their first Aeolian recording session. But the two numbers recorded on that afternoon—"Indiana" and "Ostrich Walk"—got only as far as the master disc. After hearing the playback, both sides were rejected by unanimous decision of the band.

"There it Goes Again" and "That Loving Baby of Mine" were similarly junked by the musicians. Edwards recalls that Metropolitan Opera star Florence Eastman, one of the earliest fans of the Dixieland Band, was in the studio when "That Loving Baby of Mine" was being cut. The selection would have turned out quite to their satisfaction, Edwards believes, if Miss Eastman had not been standing so close to the pickup horn when she shouted, "Oh, boys, that's good!" The record was ruined.

The Dixieland Band did, however, wax seven sides that met with their own approval and that of the recording company. These were released during the fall of 1917. "Ostrich Walk," in a faster tempo than on any other record to follow, reveals the band at its peak of precision, with Shields taking more liberties than usual on the breaks. The choruses of "Oriental Jass" are some of the finest examples of LaRocca's driving syncopation.

("Oriental Jass" was actually a number called "Sudan," but the boys didn't know it at the time and who could blame them for making up their own title?) "Reisen-weber Rag" is interesting mainly as a lesson in how to play the "Original Dixieland One-Step" without getting sued. To avoid splitting any more royalties with the copyright owners of "That Teasin' Rag" (see Chapter 7), LaRocca shrewdly improvised a new chorus for the "One-Step" and changed the title.

In describing its new series of jazz records, the 1917 Aeolian catalogue remarked: "The fascination of jazz music is in its barbaric abandon. Jazz was first introduced to the world a year ago by the Original Dixie-land Jazz Band and immediately captured the fox-trotting public with its bizarre instrumentation and unique rhythm. A full measure of the weird characteristics responsible for the remarkable vogue of this form of dance music will be found in the record selections of the Original Dixieland Jazz Band."

It is indeed unfortunate that these curious old Aeolian records, with their golden, rococco labels and unorthodox playback requirements, are so incredibly rare and that they cannot be played even when they are found. In some ways they are superior to the famous Victor series. Proper balance among the musicians was more nearly achieved, with LaRocca more to the front, and less confusing echo is in evidence. And if dynamics, or "shading," is considered new stuff to present-day dixieland bandleaders, they would do well to listen to LaRocca's treatment of "At the Jass Band Ball" (Aeolian No. 1205). Here the Dixieland Band decrescendos into the second chorus and plays it at a mere whisper, returning to the full volume of their instruments on the third chorus. The trick is executed with great feeling and finesse, and tends to relieve the monotony of three choruses played exactly the same way. Victor officials, however, did not believe that the soft chorus was suitable for dancing and insisted upon full blast from beginning to end.

In March of 1918, after an absence of more than a year, the Dixieland Band returned to the Victor studios to continue the series of dixieland classics that had begun with "Livery Stable Blues." It is interesting to note that in nearly every case, contrary to modern practice, these early jazz compositions were recorded and made available to the public in the form of phonograph records *before* they were reduced to writing. At the time of these recording sessions, the complex scores of these musical productions existed only in the minds of the musicians. Attempts at copying five-piece band arrangements note for note from the records were, in fact, totally unsuccessful, owing to the dozens of odd musical effects and subtle refinements of rhythm for which there were no known symbols. The publishers were finally content to settle for standard orchestral arrangements derived solely from the melody.

The contract between the Original Dixieland Jazz Band and Leo Feist and Company was signed on August 9, 1917, and there is evidence that Feist worked very closely with the Victor Company in the exploitation of these numbers. The sales of the Victor records were undoubtedly watched carefully by the publishers in an effort to predict future demand for sheet music and arrangements.

TABLE 5

Recordings of the Original Dixieland Jazz Band (1917–1918 Series)

RECORDING DATE	COMPANY	NUMBER	TITLE	RELEASE DATE
ca. January 30 1917	Columbia	A2297	Darktown Strutters' Ball Indiana	May 31 1917
February 26 1917	Victor	18255	Dixie Jass Band One-Step Livery Stable Blues	March 7 1917
July 29 1917	Aeolian	——	Ostrich Walk	Not Released
July 29 1917	Aeolian	——	Indiana	Not Released
August 7 1917	Aeolian	1206	Ostrich Walk Tiger Rag	*ca.* September 1917
August 7 1917	Aeolian	——	There it Goes Again	Not Released
August 7 1917 September 3 1917	Aeolian	1205	Barnyard Blues At the Jass Band Ball	*ca.* September 1917
November 9 1917	Aeolian	——	That Loving Baby of Mine	Not Released
November 9 1917	Aeolian	——	Look at 'em Doing It	Not Released

November 24 1917	Aeolian	1242	Look at 'em Doing It Reisenweber Rag	*ca.* December 1917
November 24 1917	Aeolian	1207	Oriental Jass*	*ca.* December 1917
March 18 1918	Victor	18457	At the Jazz Band Ball Ostrich Walk	June 1 1918
March 25 1918	Victor	18472	Skeleton Jangle Tiger Rag	*ca.* June 1918
June 25 1918	Victor	18483	Bluin' the Blues Sensation Rag	September 9 1918
June 25 1918	Victor	——	Mournin' Blues	Not Released
June 25 1918	Victor	18564	Fidgety Feet	September 1 1918
July 17 1918			Lazy Daddy	
July 17 1918	Victor	18513	Mournin' Blues Clarinet Marmalade Blues	February 3 1919
December 3 1918	Victor	——	Satanic Blues**	Not Released
December 3 1918	Victor	——	'Lasses Candy**	Not Released

Personnel: LaRocca (cornet), Edwards (trombone), Shields (clarinet), Ragas (piano), Sbarbaro (drums)

* correct title: Sudan

** Emile Christian in place of Eddie Edwards

These Victor records of March–July 1918 (see Table 5), although not approaching the spectacular sales volume of "Livery Stable Blues," were successful enough to saturate the market from coast to coast and may still be found lurking in the dusty corners of secondhand furniture stores.[4]

"At the Jazz Band Ball" appears to have been the most successful. The Victor version is very similar to the old Aeolian, except for a much faster tempo. LaRocca's well-known style of "hitting before the beat" is very pronounced on this disc, and his four cornet breaks certainly come under the heading of musical pyrotechnics.

"Fidgety Feet," another fast one-step, shows off the co-operation of the three wind instruments and their flawless execution of every note. In the chorus LaRocca supplies a mere framework for the ensemble—a subtle, skeleton melody with many openings for the quickly maneuvering trombone and clarinet. His famous "flying tackle" cornet breaks are the outstanding feature of this record and deserve careful attention. The four-bar introduction, like that of the "Original Dixieland One-Step," was fashioned from the snare drum "roll-off" frequently employed by parade bands. Comments the Victor record catalog of September, 1918: " 'Fidgety Feet' is a clever piece of music by LaRocca and Shields in which a novel

[4] RCA unearthed the old Victor masters a few years ago and produced a ten-inch LP (Label "X," No. LX3007) on which eight of the original twelve numbers were reissued. More recently, a group of five dedicated musicians on the West Coast have turned out an LP album called *Original Dixieland Jazz in Hi-Fi* (ABC-Paramount No. ABC-184), intended as an exact replica of the historic 1917–18 releases. All twelve numbers are included on this 12-inch disc.

rhythmic effect is produced by having two silent beats introduced here and there. The result is highly stimulating . . . You can never tell just what these jazz fellows are going to do next, but they always contrive to do something you never thought of before."

"Clarinet Marmalade" is one of the best. It was created to feature Shields, hence its title, and shows him off to good advantage in the rapid polyphonic breaks. This is the kind of record that requires several playings for full appreciation—once through, listening only to LaRocca's sharply cut notes and brilliant tone, again to catch the inspired trombone counterpoint of Edwards, and as many times as necessary to absorb the details of Larry Shield's skillful passages.

"Tiger Rag," for all its popularity, is almost never played in its entirety. By far the most involved of all La-Rocca compositions, it has nine separate and lively parts—five parts to the introduction and four variations of the chorus. Here, on this 1918 Victor, they are revealed in their original form. The "hone-ya-da" chorus, mentioned earlier in this chapter, is prominent, with the clarinet embellishment of Shields overriding the cornet-trombone "riffs." Although originally intended as a mere obbligato, subservient to the "riffs" themselves, this clarinet part became much imitated as a solo in subsequent years. The third chorus is the well-known "hold that tiger" routine in which Edwards uses his trombone to emulate the roar of the animal. The final chorus, or "home stretch," consists of collective improvisions with the cornet for the first time stepping out of its role as a melody instrument—a daring experiment in its day but quite common to modern jam sessions.

Clocking the speeds of these one-steps, it is interesting to note that "Clarinet Marmalade" is the fastest, and that four others are rendered at precisely the same tempos:

TITLE	BEATS PER MINUTE
Clarinet Marmalade	268
At the Jazz Band Ball	256
Dixie Jass Band One-Step	252
Tiger Rag	252
Sensation Rag	252
Fidgety Feet	252

Of the slower numbers, "Ostrich Walk" and "Lazy Daddy" are particularly designed for the clarinet. In "Ostrich Walk," Larry Shields takes no fewer than fifty-one breaks, and although they are of short duration, when considered collectively they seem to form a melody in themselves.

"Skeleton Jangle" is a fox trot which was built around its trombone solo, reversing the conventional procedure. This trombone solo was inspired by Franz Liszt's "Second Hungarian Rhapsody," so often played by the old concert bands in New Orleans, and bears a marked resemblance to it. The solo, played in the trombone's lowest register, is heard twice on this record. With its long, sustained tones, it is indeed a difficult passage, and any trombonist who has ever tried it will attest to the amount of wind it requires. Edwards recalls that he banged his slide against a chair, jarring the mouthpiece from his lips and causing the peculiar interruption in the first solo.

An indication that jazz was still a fad and not really appreciated is found in the following blurb in the Victor catalog for June, 1918: "A jazz band is a unique organization of which it may be said the worse it is, the better it is. If you have heard a jazz band before, and feel that you already know the worst, try this record. Yet out of the mass of sounds there emerges tunes, and as the music proceeds you get order out of chaos, and a very satisfactory order at that. One that not merely invites you, but almost forces you to dance."

But whatever the reason, for its goodness or its badness, jazz was growing ever more popular. Eighteen months had passed since the Original Dixieland Jazz Band had introduced the new music to New York City, and strangely enough no competition had yet appeared on Broadway. Columbia, in a frantic attempt at finding another genuine jazz band to compete with Victor, sent Ralph Peer, its Director of Artists and Repertoire, to New Orleans to look around. After three weeks in the Crescent City, Ralph Peer wired back:

NO JAZZ BANDS IN NEW ORLEANS

In desperation, Peer stopped off at Memphis on his return trip and brought back a colored orchestra led by trumpet player W. C. Handy. The Handy orchestra recorded half a dozen sides for Columbia, none of which commanded much public attention. Handy remained in New York and went into the music publishing business with his brother.

It was not until the arrival of Wilbur Sweatman and his "Original Jazz Band" late in 1919 that Columbia was again able to sell "jazz" records on a profitable scale.

9

Jazzing the Draft

The musical cyclone that swept New York in 1917 left a multitude of enthusiastic but somewhat bewildered dancers in its wake. The bunny hug and turkey trot fell fast into the discard as the driving, throbbing, agitating beat of jazz demanded an entirely new style of expression. There had been one-steps many years before "Fidgety Feet," "Sensation Rag," and the Original Dixieland, but here was a faster tempo and a new kind of rhythm. The old established patterns of dancing no longer filled the bill.

A few experts tried their hand at relieving the confusion. The leading organization of dance instructors known as the Inner Circle held its annual convention in New York in September and, as expected, jazz was the main topic of discussion. In an article entitled "How to Dance the Jazz," appearing in the New York *American* of November 18, 1917, they advised as follows:

> The Jazz, the most nervous of music, is wedded to one of the most erratic dances of the season.
>
> Eccentric as is the fickle, changeable, erratic dance it has the approval of the Inner Circle, the body of dancing teachers that stands for progress in the terpsichorean art in America. At its convention, held in New York in September, it recommended the Jazz as a kind of paprika of the evening's programme. The Jazz is the evening's romp.
>
> The convention selected four dances as the leaders of the

season. They were the Chinese Toddle, the fascinating drawing room dance previously described and pictured in this newspaper; the Ramble, a pleasing combination of the one-step, the
fox-trot and the waltz; the Inner Circle Tango, for which the
enthusiastic convention claimed the honor of being the aristo-
crat of all ballroom dances. As a gay little fillip for an evening's dance medley, the convention also suggested "Hello,
Pal," a one-step accompanied by the pledge of fellowship. But
the convention laid special stress upon the Jazz.

The dance was created by Oscar Duryea, who further on,
on this page, gives a correct, technical description of it. The
Jazz, Mr. Duryea, a prominent originator of dances and an in-
structor in dancing, characterizes as the "latest American fox
trot idea. . . ."

This dance is a Fox Trot, but with a peculiar rhythm
somewhat different from the usual Fox Trot—it is much
slower. The original Dixie Land "Jazz" Band consists of a
piano, cornet, trombone, clarionet and trap drums. The peculiar, somewhat discordant melody is said to be produced by
tuning each of the instruments at a different pitch; and to end
some of the strains they occasionally play what we have termed
a "crazy cadenza." Music is 4-4 tempo, but for convenience in
teaching the dance, the count of "one" has the value of 2
musical counts. Description for gentleman-partner counterpart
Waltz position throughout.

It is obvious that even after eight months of jazz
in the Big City, most listeners were still at a loss to
explain just how the "trick" was accomplished. Some
believed that the swiftly changing harmonies of the new
music were in actuality an ingenious kind of discord.
To explain the phenomenon, the ludicrous theory had
even been propounded that the instruments in the Original Dixieland Jazz Band were tuned to different pitches!
Under the influence of these popular misconceptions, is
it any wonder that jazz quickly degenerated into the
unbridled cacophony that characterized its early period?

The one-step, originally the vehicle for the quickly moving music, was fast being replaced by a new dance pattern. Jazz was becoming slower and swingier, and across the country dancers were developing new variations of the fox-trot that had swept into popularity with "Ballin' the Jack." From San Francisco's Barbary Coast came Frank Hale and Signe Patterson, a vaudeville dance team bringing with them a new western craze called the "shimmy-shewabble," later to be known simply as the "shimmy."

That Frank Hale's native Chicago was the birthplace of "shimmy" as well as of jazz appears evident from the research of various experts who, by stretching a point or two, have been able to trace this dance back as far as the Chicago Columbian Exposition of 1893. The so-called "Egyptian Dancers" who performed their scandalous "hootchy-kootchy" on the Exposition's midway may well have been the pioneers of the dance craze that was to sweep the nation almost three decades later. It seems likely that the "kooch" dance was first applied to popular music about the same time as the word "jass," and that the new version of this dance—the "shimmy-shewabble"—was actually a by-product of the first "jass" music played in that city by the Original Dixieland Jazz Band. Its westward migration evidently began with vaudeville troupes of that period, who were responsible for its immediate popularity on the Barbary Coast, and especially in what were then known as "closed dance halls."

Unlike the Charleston of the next decade, the "shimmy" was primarily an exhibition dance, limited mostly to the musical comedy stage and to the very

young, athletic, uninhibited, and—above all—daring.
Strictly a solo routine, it was performed with the toes
together, the heels apart, and the arms raised almost to
shoulder level. From that position, it was simply a
matter of putting the shoulders in motion and letting
nature take its course. The general public, however,
was content with the slower fox-trot, with its body-to-
body clutches and cheek-to-cheek intimacy.

In November of 1917, the Original Dixieland
Band formed a vaudeville act with Hale and Patterson
and opened at Keith's Colonial Theater. *Variety* reports
the success of the "jazzy syncopated" shimmy-shewabble
in its issue of November 30:

> Frank Hale and Signe Patterson haven't been dancing to-
> gether since last spring and although for several weeks they
> have been framing the present routine, it is said they opened
> "cold" at the Colonial Monday matinee. Some changes were
> effected for the second performance, probably in the way of
> eliminations. Whatever it was, the team went over with a bang
> at the night show. One thing chopped after the matinee was a
> Chinese orchestra, and at night the Chinks draped themselves
> in fancy regalia in the rear of the back hangings, sitting amid
> smoking incense. This was, however, merely during the second
> number billed as the dance "Quan Chung," a waltzy thing that
> the dancers did very prettily, garbed in creations celestial. The
> opening number was programmed as a combination of "Strut-
> ters' Ball," "Shimme-Sha-Wabble," and "Walking the Dog."
> It is a dance number of the jazzy syncopated type. Mr. Hale
> and Miss Patterson seemed imbued with the "let's go" spirit
> during the dance and their efforts quickly brought the house
> to a realization something was going on. After they exited for
> the Chinese change, the "Dixieland" (five men) orchestra took
> up the pace with a flying start. This is the same bunch stirring
> up things at Reisenweber's and they are "some" jazz players.
> Hale and Patterson were to have had two orchestras on the

stage, but after Monday night they should be satisfied that the Chinese aren't needed. The jazz players syncopated "Chinatown" for the dancers' Chinese number and it sounded very good. The boys also had two numbers alone, one being "Livery Stable Blues," and that alone planted them solidly with the house. If Hale and Patterson can carry the jazz players, their new turn is sure-fire. The final dance was the "Whirling Dervish," similar to the finish number of last season, but Hale is robed as an East Indian prince and his partner shows considerable of her slender form. Hale is accredited as one of the best of what are known as syncopated or jazz dancers and the crowd sure took to his body evolutions. The pair went over an easy hit.

The busy musicians of the Dixieland Band barely had time to pack up their instruments, dash out of the theater, and hurry over to Reisenweber's, where they began playing nightly at ten o'clock. As if this were not enough, their round-the-clock schedule also included regular Sunday night concerts at the Winter Garden during the season of 1917–18, appearing on the same bill with Jack Norworth, Frank Fay, Ed Wynn, Chic Sale, and Fred Astaire. During the week, at this same theater, Al Jolson was electrifying the crowds in *Sinbad*.

Then there were the private engagements forced into an already overcrowded schedule—everything from a senior dance at a girl's high school in Mamaroneck, Long Island, on May 1, 1918, to the fabulous private parties of Al Jolson. The band played at three of the Jolson events that year—one on St. Valentine's Day, another on March 23, and still another on June 1. These parties, held at the various homes of Jolson's friends on Long Island, were lavish affairs. LaRocca recalls that "champagne flowed like water" and beautiful girls abounded by the score. Nick, however, was a teetotaller, and he adds with a chuckle that "this was where I had the advantage on the boys."

In May of 1918 a new Liberty Bond drive was launched and the Original Dixieland Band toured the city on the back of a truck, sending the strains of dixieland jazz echoing up and down Broadway. LaRocca claims that more than a million dollars in bonds was sold from the tailgate of the truck in a single afternoon.

Military conscription was also gaining momentum, and now the Original Dixieland Band was beginning to have its troubles. Draft notices were forwarded from New Orleans and the Dixieland musicians (except for Tony Sbarbaro, who was under draft age) were summoned for physical examinations. Because the very life of the Original Dixieland Jazz Band depended upon maintaining the original combination of musicians, its fate hung in the balance. Several hundred numbers had been tirelessly arranged, rehearsed, and memorized by this combination of five men. Their situation was not that of the average note-reading orchestra, where substitutions in personnel could easily be made. In the case of the Dixieland Band, the loss of a single man—any one of the five—could have proved disastrous. It could, in fact, have resulted in the total collapse of the organization at the peak of its career. And industrial deferments were almost unheard of in the first World War.

In compliance with his notice, Nick LaRocca appeared one afternoon at the local draft board for Division No. 158, New York City. He recalls the calamity caused by the twitch in his left shoulder, a nervous disorder possibly resulting from the long and strenuous hours of a jazz musician. The applicants had stripped and were standing in a long jagged line awaiting their medical examination, when Nick's dancing shoulder began its erratic movements. One of the doctors, who had been watching this phenomenon from across the room, came over, pulled the wavy-haired cornetist out of line, and had him stand in a corner, where the others could not see him.

"Do you *always* do that?" demanded the doctor.

"Do what?' asked Nick.

"This," said the doctor, jerking his shoulder.

Nick went on to explain he had been under treatment many months for this nervousness with little success. Another doctor came over and expressed the opinion that the whole thing was an act and demanded the name of the specialist who had been treating him. Other physicians joined the consultation until LaRocca was completely surrounded by doctors, no two agreeing on the exact meaning of the symptom. Finally one of them waved LaRocca back to the cloakroom.

"Get out of here," he ordered, "before you have us all doing it."

The bandleader was classified 4-F because of the malady, but the federal agents tailed him for a few weeks just to make sure the twitch stayed with him. These investigators could be seen sitting at a table in Reisenweber's, in the Greek restaurant across the Circle, where LaRocca usually ate, and sometimes in the first row of the Colonial Theater.[1] LaRocca's ailment remained with him during his lifetime but seemed directly

[1] From the author's interview with Nick LaRocca, July, 1957.

connected with his musical expression, for it rapidly diminished after he quit the entertainment business.

Larry Shields was likewise deferred from the draft because of a highly sensitive nervous system that caused him many sleepless nights. While the Original Dixieland Band was playing at Reisenweber's, Edwards, Shields, and Sbarbaro rented a room in the Reisenweber Building. Edwards recalls the nightmares of Shields and remembers especially one particular night shortly after the clarinetist had received his draft notice. Larry woke up screaming, "They're coming after us!" The jocular Edwards, disturbed from a sound sleep, shouted back, "That's right. Better have your musket ready!"

But although LaRocca and Shields had been rejected by the medical board and Sbarbaro was too young to worry, the troubles of the Dixieland Band had not yet ended. Just when the jazzmen considered themselves safely through the storm, their ship was upset by a mighty blow. On July 20, 1918, Eddie Edwards was drafted.

Although the first reaction was one of panic, the musicians quickly adjusted to their problem. It may have been Nick LaRocca's steadfast determination and unwavering confidence in the Original Dixieland Jazz Band that kept it from breaking up during this critical hour. Edwards played with the band until the date of his induction; on July 30, 1918, when the trombonist marched off to Camp Upton, New York, to join the 152nd Depot Brigade of the famous Rainbow Division, the four remaining members boarded a train for New Orleans in search of a substitute. It is interesting to note that although the Original Dixieland Band at this time was the highest-paid dance orchestra in the world and could have its pick of almost any trombonist on earth, the members were forced to return to their home town for the replacement.

Then a most amazing, almost unbelievable thing happened. Instead of accepting one of the many dozen supposedly competent tailgate men who had been clamoring for the job in the Crescent City, the Original Dixieland Jazz Band selected a cornet player and taught him to play trombone!

Emile "Boot-mouth" Christian, a cornet player of no mean ability who had played with many small combinations in New Orleans and Chicago, now took over the trombone chair in the band he had for so long idolized. Emile had been a close buddy to the members of the Original Dixieland Band many years in advance of their northward expedition in 1916, had cheered them from the sidelines during their great success in New York, and it is likely to assume that he had thrown his hat into the ring long before any substitutes were needed.

For five solid weeks the Original Dixieland Jazz Band practiced at Larry Shields' house in New Orleans before returning to the capital of jazz in the North. As Christian had played slide cornet and already knew the slide positions of a trombone, it was only necessary for him to develop an embouchure for the larger mouthpiece. Day after day, the Victor records were played over and over on a wind-up phonograph while the industrious

Emile studied the trombone parts created by his pal, Eddie Edwards. The Columbia records that Emile waxed with this band in England a year later reveal that he copied these classic countermelodies with incredible accuracy. None but the most studied of experts could detect that the trombonist was Christian and not Edwards. As for Christian's own creative ability, the counterpoint he developed on new numbers established him as a tailgate man second only to Edwards. Possessing a most remarkable ear, he was able to learn a band's whole repertoire within the short period of a few weeks.

The Original Dixieland Jazz Band returned to New York just before the Labor Day week-end, all members riding in the same railway coach. LaRocca advanced Emile train fare and living expenses pending the band's return to regular salary. On September 7, 1918, the new combination reopened at Reisenweber's and also continued its vaudeville act with Hale and Patterson on the Keith circuit.

The war ended with an armistice on November 11 and a jubilant America was about to enter a carefree and energetic era that took its name from the very music that characterized it—the Jazz Age.

For a while it seemed that the band's troubles were over. Emile fitted into the combination as if he had always belonged, and there were no Yankee accents on the bandstand to disillusion the fans at Reisenweber's. On November 23 a British producer named Albert deCourville signed the Original Dixieland Band for a ten-week engagement at London's Hippodrome, to commence on or about March 2, 1919, at a salary of 220 pounds per week—then the equivalent of $1,056 in American money.

But the greatest tragedy of all was about to strike, an unexpected turn of fate that threatened to crush the band at the peak of its popularity.

10

The Silent Bandstand

By late autumn of 1918 the great influenza pandemic that was sweeping the world had sent many millions of Americans to bed with aching bodies and high fever. Nearly a half million of the more unfortunate had already died from its complications. Although the healthy young musicians of the Original Dixieland Jazz Band had survived the first two waves of the dread disease, the last and comparatively minor epidemic of that winter was about to take its fatal toll. Henry Ragas, whose physical condition had been severely undermined by late hours and an excess of alcohol, was an easy target for influenza.

Henry's decline had begun the previous summer. Plagued with family problems that drove him to distraction, he quickly turned to drink as the only escape. In November, realizing the gravity of Henry's state of mind, LaRocca had sent him home to New Orleans in hopes that he could straighten out his personal affairs. But when the anguished pianist returned to the band early in December, matters had grown only worse, and as he lost himself in alcohol his absenteeism from the bandstand became more frequent and of increasing duration. A failing appetite and lack of nourishment brought on the long chain of miscellaneous illnesses, the inevitable "flu," and now pneumonia. The other members of the band contributed to Henry's expenses during his long period of unemployment, but medicine, doctor bills, room rent, meals, and a hundred debts had left the dying pianist totally destitute.

His fellow musicians, who had stayed away during the flu contagion (or because of the vicious bulldog he kept in his hotel room), now called regularly at Ragas' bedside. LaRocca remembers that the despondent pianist had lost all desire to live.

The end came at Bellevue Hospital around eight o'clock on the cold and gloomy evening of February 18, 1919—just two days before the band was to sail for England. A few days later the following black-bordered memorium to the twenty-eight-year-old musician appeared simultaneously in all New York newspapers:

In fond remembrance of our dear pal and pianist,
H. W. Ragas, who departed from this world
Feb. 18, 1919. May his soul rest in peace.
The Dixieland Jazz Band

The loss of Henry Ragas was a tragedy of major proportions to the Original Dixieland Jazz Band for professional as well as personal reasons. The rhythm of the new jazz music had been the hardest thing to imitate, and in this respect the piano was the most important instrument. Although he had never taken a lesson in his life and could not read music, Ragas was beyond doubt one of the greatest ensemble pianists in the history of jazz. His "heavy" left hand provided a strong chord foundation for the structure of the music, while his right added another melodic voice to the band. Unlike most jazz pianists of today, Henry didn't play just rhythm alone with an occasional embellishment. Here was a "ten-fingered" pianist whose style contributed body and background strength to the ensemble.

Developing a substitute pianist in time for the March 2 debut at London's Hippodrome was out of the question. LaRocca wired deCourville immediately, requesting a month's extension, which was granted.

A local pianist by the name of Willy Hollander had filled in during the illness of Ragas. When it finally appeared that Henry would never recover, LaRocca sent to New Orleans for Eddie Shields, one of Larry's younger brothers. Eddie arrived without delay, anxiously cracking his knuckles and bursting with joy over his big opportunity. But even Eddie couldn't "catch" the rhythm. Two weeks later Eddie was on his way home and LaRocca was scouring New York for a pianist who could match their style. Sidney Lancefield (now a Hollywood movie executive) was tried next and came somewhat closer but wasn't quite right for the job. Publishers sent over song "pluggers" by the score, and each one sat in with the band for a chorus, but none could play the rhythm.

After a large number of pianists had tried and failed, LaRocca was for the first time genuinely worried over the band's future. He began to feel that the London job was a jinx and even tried to have it cancelled. Bob Simons had offered the band a long-term engagement at the Hotel Martinique in Atlantic City. But William Morris, the New York agent through which the deCourville contract had been arranged, threatened an injunction if the band did not follow through with its London commitments.

One afternoon, while Nick sat unhappily in his lonely room at the Pontchartrain, nervously fingering the valves of his cold cornet, his reverie was interrupted by an unfamiliar knock on the door. The caller, a dapper gentleman with a neatly trimmed moustache, introduced himself as J. Russel Robinson. Word of the band's predicament had been passed up and down publishers' row, and Robinson, who was manager for W. C. Handy's publishing firm, was anxious for a crack at the job. Later that day he sat in with the band at a practice session and amazed the members by playing a solid piano part behind every one of their original numbers.

Robinson's adaptability to the style of the Dixieland Band was no accident of nature. Although he was born in Indianapolis on July 8, 1892, most of his youth was spent in the South. Feeling the wanderlust at an early age, he and his brother, the late John C. Robinson, formed a two-man piano and drum team and played in silent movie theaters in the larger cities throughout the South, beginning in Macon, Georgia, in 1908, and spanning the Gulf to New Orleans. On Canal Street they played for the Herman Fitchenberg Enterprises, a company operating the Wonderland, Dream World, Princess, and Alamo theaters. Russ claims that they were known as the hottest piano and drum team in that city. In 1909 his first composition, "Sapho Rag," was only the first of more than six hundred published works to follow, including such perennial favorites as "Margie," "Mary Lou," and "Singin' the Blues."

Russ and his brother played along with the first

phonograph records of the Dixieland Band when these discs began flooding the South in 1917, picking up the new rhythm and learning the original Dixieland compositions as they played through their daily routines at Canal Street nickelodeons. When Russel, well imbued with the Dixieland spirit, returned north a few years later to cut piano rolls for the QRS Company, it was only natural he choose the "Original Dixieland One-Step" (QRS No. 100800) as his first number. At the time he joined the Original Dixieland Jazz Band in 1919 he had cut hundreds of rolls for this company and was already known throughout the country for his outstanding work in this field.

Although the Dixieland Band played its last engagement at Reisenweber's in the latter part of January, 1919, the band continued to rehearse there for another month. With Robinson on piano they finished their vaudeville stint on the Keith circuit and made last-minute preparations for their historic misson of carrying jazz across the seas to Europe.

11

London: Spreading the Gospel

Larry Shields bounded up the stairs of the Pontchartrain Hotel four steps at a time, so anxious was he to break the news. On the third floor landing he spotted Nick LaRocca, spun him around a few times in an awkward waltz pivot, and blurted his glad tidings: he and Clara had just been married.

"Buy me another ticket," exulted the lanky clarinetist. "I'm takin' my family along."

Four years earlier in Chicago, when Larry was playing with Tom Brown's band, he had first met the girl of his dreams. Lamb's Café was then one of the Windy City's most popular spots for dancing, and Clara Belle Ferguson and her beaux had come there often to turkey-trot to the lively tempos of the Band from Dixieland. It was here that Clara first became acquainted with the New Orleans clarinetist; and now, just a few weeks before the Original Dixieland Jazz Band was to sail for England, they had become man and wife.

The trip to England was, therefore, more of a honeymoon than a job for Larry—not that playing the clarinet was ever really work for him to begin with. On March 22, 1919, he and Clara waved farewell to the Manhattan skyline as the *R.M.S. Adriatic* steamed out of New York harbor carrying the members of the now

world-famous jazz band. On the passenger list were:
"D. Jas. LaRocca, Emile Christian, Anthony Sbarbaro,
Mr. and Mrs. J. Russel Robinson, and Mr. and Mrs.
Lawrence Shields."

Shipboard parties and concerts kept the musicians
in trim, and on April 1 the hoarse blasts of the *Adriatic's*
foghorn announced to Britain that the American emis-
saries had at long last arrived, bringing to the Old World
its first taste of the fast new music from across the seas.
Disembarking at Liverpool, they travelled by train and
arrived that same evening in London. Late at night the
five musicians reported directly to the Hippodrome
Theater, where producer Albert deCourville greeted
them with a strange mixture of delight and apology. It
seems that American doughboys and government em-
ployees occupied every spare bed in the city of London.
The jazzmen would have to spend their first night in
the chorus girls' dressing room.

The next morning British newspaper reporters
crowded the dressing room doorway. What is jazz? What
does it sound like? What does the word mean? The
London *Daily News* of April 4, 1919, reported as follows:

. . . as to the word "jazz," the bandsmen rejected both the current explanations. They will not have it that the word is of Red
Indian origin, or that "jazz so" is a term of praise in the
dialect of the Negroes in the southern states. The word was invented by someone in Chicago . . . it is possibly a purely
onomatopoeic expression . . . In view of the unkind and dis-
respectful things which have been said about Red Indians and
Negroids and West African savages, it should be stated that
the players are all white—as white as they can possibly be.

The five American musicians warmed up on the stage as a terrified group of British journalists huddled together in a far corner of the theater. Their reactions to the sounds of jazz are typified in this release in the April 10 issue of *The Performer*:

I had an experience the other evening. Whether I'm to be envied or not depends on personal tastes . . . a semiprivate exposition of real jazz by the Original Dixieland Jazz Band. At least I was told by Mr. deCourville that this was THE Original Dixieland Jazz Band, and from the noise "kicked up" . . . I may well believe him . . . I am assured that there are only two original jazz bands in America. Why two, I cannot say . . . I'm told that the other of the two original bands is but an imitation; a fact which seems to clear the atmosphere somewhat. Then I'm told that the Dixieland lot are the original, so it seems that poor America has to be content for the nonce with a mere substitute . . .

The Original Dixieland Jazz Band opened at the Hippodrome in the musical review "Joy Bells" on April 7, appearing in a specially staged café scene. The ovation following their first number was deafening, due in large part to the number of American doughboys scattered throughout the audience. Many of these had been fans of the Dixieland Band before the war, had heard them on the first Victor records, and now came to cheer and applaud their idols. The fever spread through the theater until every last man and woman was on his feet, shouting and clapping in a manner peculiarly un-British. That night when the curtain came down, George Roby, the star comedian of the show, approached deCourville in a seething rage and served his ultimatum: Roby or the jazz band would have to go— deCourville could have his pick. And so it was that the

Original Dixieland Jazz Band was permanently removed from the cast of "Joy Bells" after an engagement that lasted exactly one night.

Town Topics, in its issue of April 12, conjured up an explanation of the mysterious disappearance:

The Dixieland Jazz Band appeared in "Joy Bells" at the Hippodrome last Monday but since has been withdrawn, presumably on account of that ubiquitous complaint, influenza. On the occasion of their performance, they gave us a demonstration of undiluted jazz, and it must be admitted, despite all that has been thought and said to the contrary, there was a certain charm in the mournful refrains, dramatically broken by cheery jingles and a miscellany of noises such as one generally hears "off."

Especially fascinating to *Town Topics* was the use of solo "breaks" during ensemble choruses, as this reporter goes on to explain:

At one moment, the whole orchestra would down tools while one member tootled merrily or eerily on his own account, and the whole would resume again, always ready to give a fair hearing to any other individual player who suddenly developed a "stunt." The conductor was most urbane about it all, but everybody was perfectly happy, not excluding the audience who appreciated a novelty not unartistic.

Tony Sbarbaro recalls with a chuckle that every British orchestra at that time used two drummers, one for the bass drum and one for the snare. It is easy to understand, then, how the London *Daily News* seemed so obsessed with Tony's drum installation: "The trap drummer who plays the big drum with his feet and a side drum, the cymbals, and heaven knows what besides, is the most important man of them all . . ."

DeCourville was not worried in the least about getting stuck with a jazz band. Newspaper reports, a near-riot at the Hippodrome, withdrawal of the band from "Joy Bells," all had served to arouse public interest. Thousands sought the oportunity to hear this strange music firsthand, if only out of curiosity, and few London café owners were unwilling to give it a try. The Original Dixieland Jazz Band opened at the Palladium on April 12, drawing this wry comment from *The Star*, April 19:

It is an interesting study to watch the faces of the dancers at the Palladium when the Original Dixieland Jazz Band, which is said to be the only one of its kind in the world, is doing its best to murder music. Most are obviously bewildered by the weird discords, but some, to judge by their cynical smiles, evidently think that it is a musical joke that is hardly worth while attempting. Perhaps they are right.

Weekly periodicals continued to battle one another over the relative merits of jazz, witness this quote-within-a-quote from *Pall Mall*, April 23:

" 'The Original Dixieland Jazz Band has arrived in London,' says an evening paper. We are grateful for the warning." —Punch

I can't help thinking Mr. Punch left the sentence unfinished; it should have continued, surely: "for the warning to go to the Palladium early and book seats."

Despite the relentless attack on them by the London press, the Dixieland Band became increasingly popular. Alternating between the Palladium and a theater in Glasgow, Scotland, they pulled down the American equivalent of $1,800 per week. There was standing room only wherever they played.

A distinguished patron haunted the Dixieland Band wherever they went. Lord Donegall, a close friend of the Prince of Wales, became fanatically interested in their music and even arranged a command performance before King George. LaRocca remembers passing through three different guards before being admitted to the courtyard. The curious musicians were carefully scrutinized by the gathering of British nobility who stared through their lorgnettes, according to LaRocca, "as though there were bugs on us."

After Tony had nervously set up his drums and Russ Robinson had explored the keyboard of the ornate grand piano, LaRocca stamped his foot twice and the little group exploded into its steaming version of "Tiger Rag," the original composition that had so successfully introduced them to Chicago, New York, and London. The royal audience, perhaps having expected a polite form of chamber music, appeared petrified at the onset, and a few members of the court glanced about uneasily for the nearest exit. At the conclusion of the number, after an embarrassing silence, his majesty laughed his approval and began to applaud energetically, followed respectfully by his loyal but bady frightened entourage. The encore, "Ostrich Walk," was received with somewhat less tension.

Quick to take advantage of the jazz boom, England's Columbia Graphophone Company prevailed upon the Americans to wax four of their famous compositions —"At the Jazz Band Ball," "Barnyard Blues," "Ostrich Walk," and "Sensation Rag" (Columbia Nos. 735 and 736). On April 16, the day of the first recording session, company officials were so overcome by the agitating beat of jazz that they signed the band for another seven platters (See Table 6).

The last five of this series, including titles such as "Mammy O'Mine" and "I Lost My Heart in Dixieland," are a real "find" for connoisseurs of good dixieland jazz, for they are the only existing recorded examples of how the Original Dixieland Band actually sounded on standard dance tunes of the day, numbers more commonly played by the band for dancing. As such, they are much slower and more relaxed. It is a nostalgic, plaintive, melodic kind of jazz—perhaps "melodic" describes it more closely than any other word. Better balance is achieved among the wind instruments, as compared with the Victor series, with LaRocca's inimitable syncopated cornet driving steadily but with amazing ease through the well-composed ensembles.

Emile Christian's counterpoint is always in good taste, and the work he does in "Satanic Blues" ranks with the best of Edwards. The waltz numbers, such as "Alice Blue Gown," prove that LaRocca's distinctive habit of rushing the beat—the very root of his syncopation—was not limited to four-four time, and that waltzes actually can be jazzed without altering their basic time. Unfortunately, his brilliant syncopation in the choruses of " 'Lasses Candy" is largely buried under the prominent clarinet-trombone counterpoint, but when these "licks" come to the surface they reveal a kind of syncopated phrasing duplicated by no other cornetist. The band again recorded what was called "Oriental Jass" on the Aeolian disc, this time under its correct title,

TABLE 6

*Recordings of the Original Dixieland Jazz Band
(1919–1920 Series, England)*

RECORDING DATE		COMPANY	NUM-BER	TITLE
April 16	1919	Columbia	735	At the Jazz Band Ball*
				Barnyard Blues**
May 12	1919	Columbia	736	Ostrich Walk*
				Sensation Rag**
May 19	1919	Columbia	748	Tiger Rag**
				Look at 'em Doing it*
August 13	1919	Columbia	759	Satanic Blues*
				'Lasses Candy*
January 8	1920	Columbia	804	Tell Me*
				Mammy O'Mine*
January 8	1920	Columbia	805	My Baby's Arms**
				I'm Forever Blowing Bubbles (waltz)**
January 8	1920	Columbia	815	I've Got My Captain Working For Me Now**
				I Lost My Heart in Dixieland*
May 14	1920	Columbia	824	Alice Blue Gown (waltz)**
				Sphinx**
May 14	1920	Columbia	829	Sudan**

* Reissued on English Columbia No. 33S 1087
** Reissued on English Columbia No. 33S 1133

Personnel:
Cornet: Nick LaRocca
Clarinet: Larry Shields
Trombone: Emile Christian
Drums: Tony Sbarbaro
Piano: (first four records): J. Russel Robinson
       (last five records): Billy Jones

"Sudan." LaRocca explains that the tune was taught them by Frank Hale of the Hale & Patterson team for accompaniment to one of their oriental dance routines, and that they never knew its right title until they arrived in England.

Fortunately, all seventeen sides of this series have been revived by the Columbia Graphophone Company in their two recent LP reissues, *The Original Dixieland Jazz Band in England 1919/1920*, Volume 1 (No. 33S 1087) and Volume 2 (No. 33S 1133). Excellent photographs of the band with Billy Jones on piano grace the front covers, backed up with well-worded commentaries by Brian Rust. In his commentary for Volume 1, Mr. Rust expresses the feelings of many:

> Of course, not everyone liked, or even approved of the apparently crazy new music; it was, and to some extent still is, the target for a great deal of ill-informed criticism and diatribe from the pulpit, the Press and even politicians. But it survives; it has changed somewhat with the passage of years, as any art form must, but the earliest genuine jazz performances, caught for us by the crude recording apparatus of those days, still sound as fresh and exhilerating as when they first bloomed. Derided during the 'thirties, when everything was for Swing and elaborate arrangements, original Dixieland music is now accepted by young and old alike, the latter perhaps with feelings of acute nostalgia, the former in the same spirit of innocent pleasure as their forebears of 1920. We have all found that, so far from being stiff and old-fashioned, the Dixieland rhythms are much more supple than the metronomic chugging that passed for "swing" just before and during the second World War, and the melodies had real tunes, that were not weary repetitions of threadbare phrases.

The Dixieland Band closed at the Palladium on April 26 and opened two days later at the Martan Club at 6-8 Old Bond Street. The reporter for the *Sunday Evening Telegram* expressed his carefree views in the issue of May 4, 1919:

> . . . It is sure some band to dance to, though I'm not sure its good for the digestion to eat to it! Yet I had to eat, because the cakes were so topping. Every sort of fox-trot and jazz there ever was seemed to be danced, but thank goodness, the only shimmy shake was done by the pianist. All the tables were taken quite early in the afternoon, so it looks as if "Martans'" has caught on again.

Martans' had indeed caught on, and the club became so identified with the raucous little band of American rebels that the management quickly changed its name to the Dixie Club. The band continued to create its musical disturbance there for two months, alternating with a four-piece tango band led by the red-haired English pianist, Billy Jones.[1] Jones was later to replace J. Russel Robinson in the Original Dixieland Jazz Band.

Robinson, approaching his zenith as a popular composer, turned out several hits for British consumption. The most successful of these was a catchy tune called "Pip-Pip, Toot-Toot, 'Bye-'Bye" that had Englishmen singing in every pub and music hall in London. This was followed by three show numbers for Beatrice Lillie—"Bran Pie," "Picadilly Jim," and "Come Along Mary," the latter being the first song ever sung by Miss Lillie outside the chorus line.

Following the signing of the Treaty of Versailles on June 28, 1919, a vast array of royalty, dignitaries, and officers of high rank gathered at London's Savoy

---

[1] Billy Jones (piano), Grimshaw brothers (banjos), Chris Lee (drums).

Hotel for the great Victory Ball. Among those present were King George and the royal family, Marshal Foch and General Pétain, General Pershing and his staff, the American ambassadors, and all the crowned heads of Europe. The 150-piece Marine Corps Band played for the ceremonies preceding the celebration. On hand to provide dance music was the Original Dixieland Jazz Band. While the teddy bear atop Tony Sbarbaro's huge bass drum waved a miniature American flag, the New Orleans musicians opened the dance program with the "Star Spangled Banner," astounding the multitude of guests by playing nearly as loud as the much larger Marine Band. When they stopped playing, cries of "Bully!" and "Viva!" echoed from the gaily decorated rafters of the hall. With the house now firmly on their side, the Dixielanders immediately stomped off with a steaming version of "Tiger Rag." Soon the dance floor was a tossing sea of bright-colored uniforms, lavish evening gowns, and glittering jewelry, as dignified people of a dozen nationalities explored their talents for doing the American one-step. Meanwhile, a squad of Marine Corps musicians gathered around the bandstand to find out how so few men could play so much music.

Before the expiration of the deCourville contract, LaRocca negotiated a deal with the agency of Mitchell and Booker, who in turn secured a job for the Dixieland Band at Rector's, 31 Tottenham Court Road, beginning June 29. The band had its longest stay at this club, playing for crowds that grew constantly in size until a larger place was called for. Mitchell and Booker then arranged the engagement at the Palais de Danse, a huge dance hall at Brook Green, Hammersmith. Robinson, however, didn't like the move, felt that the band was losing it's prestige, and walked out on October 11. That same night he was replaced by Billy Jones, the British pianist who had been studying the jazz piano style very assiduously during the four months his own band had been alternating with the Dixieland at Rector's.

That a larger hall was demanded to accommodate the growing number of jazz fans in England is borne out in the gate receipts for opening night, when the Palais de Danse boasted the amazing total of 5,800 paid admissions. The whole character of the Palais soon changed due to the new music, as an entirely different clientele replaced the original dance hall crowd. On November 22 the Palais de Danse opened as a night club, complete with tables, drinks, waiters, and brand-new decorations. In its first issue, dated April, 1920, the *Palais Dancing News* published an interview with Nick LaRocca in which he is purported to have claimed that "jazz is the assassination, the murdering, the slaying of syncopation. In fact, it is a revolution in this kind of music . . . I even go so far as to confess we are musical anarchists. . . our prodigious outbursts are seldom consistent, every number played by us eclipsing in originality and effect our previous performance." The editor, having thus skilfully translated LaRocca's American slang into King's English of incredible polish, went on to proclaim that the members of the Original Dixieland Jazz Band were "musical geniuses. . . for they are unable to read from music, and play from memory, thus concentrating their attention on producing and inventing new, but appropriate and much appreciated, dance accompaniments." Such an insight into the true nature of jazz was indeed advanced for this early period.

Mitchell and Booker were optimistic enough to exercise the six-month option clause in their contract, securing the band's engagement at the Palais until June 26, 1920.

A short vacation following the Palais job gave Larry Shields and his bride their long-awaited chance for a honeymoon in France. Clara had been studying French for a whole month in preparation. But her newly acquired knowledge was never put to use, for they were escorted all over Paris for two weeks by an old friend of Larry's—a French juggler he had met in vaudeville many years before.

Meanwhile, back in London, the dashing, wavy-haired leader of the Original Dixieland Jazz Band continued to "make out" like a best-selling novelist at a garden club tea party. Reports that the cornetist romanced with nearly every girl in London are grossly exaggerated, as he had time to meet only half of them. In fact, the tour of the Dixieland Band in England ended on a slightly sour note when an enraged Lord Harrington, father of one of London's most beautiful debutantes, chased the band down to the docks at Southampton with a loaded shotgun in his trembling hands. Years later a French magazine writer, delving into the mysterious past of the famous jazz band, inquired of its leader:

"Why came you never to Paris?"

LaRocca replied, "I was lucky to get out of England alive!"

The Dixieland Band, safely aboard the *S. S. Finland*, left England at 8:40 P.M. on July 8, 1920, and arrived in New York Harbor nine days later. After a day of quarantine, they disembarked on the eighteenth. With lucrative job offers coming from all directions—and all of them demanding nothing less than the original combination of personnel—LaRocca quickly sent a cablegram to Shields, urging him to return from Paris on the next boat. The clarinetist and his wife rushed back to the Regent Palace Hotel in London to pick up their luggage and immediately booked passage on the next ship out of Southampton—the *S. S. Vestris*.[2]

Emile Christian, finding Edwards out of the army, decided to return to England to capitalize on the recent success of the band. Upon his return to London, Emile tried desperately to put a band together for a job Mitchell and Booker had lined up for him in Paris. Except for Billy Jones, English musicians hadn't yet caught on to the new rhythm and Emile's only hope was to entice three or four American jazzmen over to the continent to complete the combination. Emile's letter to Nick tells the story:

London September 6, 1920

Friend Joe:

Well, son, I arrived without any trouble. I had a talk with Mitchell and he asked me if you boys were coming back. I told him yes and he asked when. He said, "Do you think they're making so much money over there?" and I said yes, they're making more. He said, "What do you call more?" I said $200 a man and he all but called me a liar . . .

---

[2] The printed program for a concert aboard the *Vestris*, dated July 24, 1920, gives a "Mr. L. Shields" as the announcer for a presentation of humorous songs, Scottish anecdotes, a violin solo, a duet, and "a little ragtime." The program notes overlooked the fact that the quiet, unassuming Mr. L. Shields also played clarinet in the ship's orchestra.

Well, Joe, I guess we'll go to Paris as soon as I can get my combination together. Mitchell asked me if I could get the men over. I said yes, LaRocca told me if I wanted anybody to let him know, and Mitchell said, "Well, I guess LaRocca will want a commission." So, from what I see there's no chance of getting a commission.

So please do me the favor and help Stein and the men over for me . . . I would like Philip Napoleon on cornet and Johnny Costello on clarinet . . . And Joe, take a tip from me, from what I understand Old Man Harrington is going to fill you full of bullets if you ever put your foot in England again. So be careful and watch yourself if you ever intend to come back to England.

Oh boy, I got the *Sphinx* we made on the record and I must say the more I play it the better it sounds!

Your pal,
Emile

But the situation was fraught with problems, as evidenced by LaRocca's reply of October 3, 1920:

308 West 58th St. October 3, 1920

Dear Emile:

. . . I have seen those boys, but what can I do? It takes money to start things, and furthermore . . . the cornet player will go over providing he has his own band, or is willing to use you on trombone . . . How are we to get these boys over when they cannot be shown where there is work? You can hardly expect anyone to put out the money for passports, etc., without a contract and the necessary fares advanced or paid by Mitchell.

. . . I can see you are stuck for men. . . What I would do if I were you is to get your band started with men you can pick up over there, as you know just why I left on the same account. I sat down many evenings to think of a way of getting men over but finally came to the conclusion that I was losing time . . .

I understand Stein is going with Bea Palmer's act, so that will leave him out . . . Giardina has gone back to Chicago, Sharkey to New Orleans, and now the only boys left are the

Napoleon bunch, and they seem to want to go together as a band. But from the cornet player they expect to get big jack, and you say Mitchell will not pay the money. . .

Just received the English records and let Edwards hear them, and he said they were great . . . They are the best we made in England. You can hear every instrument plainly, and at that the clarinet is too loud, but they are the best we made in London.

Nick

So the plans for another American jazz band in England fell flat as Emile joined an outfit called the Broadway Sextette, which played around London for about a year. In 1921 he left Britain to play with the American Five at the Frolics Café in Paris and still later at Zelli's Cabaret in the Montmarte district. After a year with Harry Boschart's ten-man dance band in Germany and other parts of central Europe, Emile returned to Zelli's. Then, from 1926 to 1930, he held a trombone job with Lud Gluskin's famous fifteen-piece orchestra, which had a virtual monopoly on the top jobs on the Continent. Travels through India and the Middle East occupied his later years and it wasn't until 1939— twenty years after he had left for Europe with the Original Dixieland Jazz Band—that Emile Christian returned to his native New Orleans. At the time of this writing he is plunking a terrific bull fiddle in Roy Liberto's Dixieland outfit at the Famous Door on Bourbon Street.

12

At the Threshold of the Roaring Twenties

Significant changes had taken place in American jazz while the Original Dixieland Band was touring England. Tin Pan Alley had monopolized the field and was turning out jazz tunes in mass production, just as it had done earlier with ragtime. Singers like Al Jolson, Billy Murray, Arthur Fields, Frank Crumit, and Sophie Tucker had become an important part of the scene. Almost every jazz band had been enlarged to include a saxophone and sometimes a banjo, or even a tuba. But jazz was still hot, still popular, and still played by ear.

Broadway had not forgotten the quintet of New Orleans boys that had started this musical swampfire nearly four years earlier. When the *S. S. Finland* steamed into New York Harbor on July 17, 1920, carrying the members of the Original Dixieland Band, booking agents were waiting on the docks.

J. Russel Robinson, who had been playing society dances and some of the better hotel jobs with the orchestras of Mike Markel, Meyer Davis, and Harold Stern since his return to the States, now was prevailed upon to rejoin the old combination. Back in the trombone chair was Eddie Edwards, who had been discharged from military service in March, 1919, just a month after the band departed for England. At that time he had formed his own jazz band, turning again to New Orleans for raw material. Sharkey Bonano, the diminutive cornetist, who always wore a brown derby when he wasn't hanging it on the end of his horn, carried the melody, with Tony Giardina adding clarinet licks. Oscar Herman sat in on piano, and the mysteriously recurring Johnny Stein, who had brought the nucleus of the Original Dixieland Band from New Orleans in 1916, gave the combination that solid dixieland beat. With the daddy of all tailgate men, Eddie Edwards, on trombone, it must have been one of the greatest outfits in New York at that time. They played several successful months on the Keith circuit during 1919 and cut a few sides for Victor, none of which was ever issued.[1] The following summer, at the College Club on Coney Island, Edwards was playing trombone with Jimmy Durante's band, much to the displeasure of the club's manager, who demanded to know where Durante ever got "that high-priced guy."

Eddie and Russ now rejoined their buddies to complete the old line-up as the Original Dixieland Jazz Band began rehearsing for its September 25 opening date at the Folies Bergère, a new night spot (formerly the Bal Tabarin) over the Winter Garden Theater. A few blocks away at the Palais Royal Café, a relatively unknown violinist named Paul Whiteman, a huge man with big ideas, was leading a small jazz orchestra that included such old hands as Gus Mueller, Tom Brown's former clarinetist, and composer of "Wang Wang Blues." Gussie had long ago swapped jobs with Larry Shields in

[1] "Dardanella," "Castle of Dreams," and "I Might Be Your 'Once-in-a-while'."

Chicago, had put in a stretch with Bert Kelly, and now was contributing his reedy trills to the Whiteman effort.

But the crowds still flocked to the Folies to be driven into madness by the tantalizing syncopation of the Original Dixieland Jazz Band. Featured with the band on opening night and widely advertised in the New York press was a vivacious blonde whose provocative body gyrations had begun to attract the attention of Broadway impresarios. Her name was Maryanna Michalski, but she was billed as Gilda Gray, "The Sweetheart of Shimmy." Although Hale and Patterson had introduced the shimmy to New Yorkers four years earlier, it was undoubtedly the electro-dynamic personality of Gilda Gray that made the nation shimmy-conscious, began a wave of moral protests, and thereby carved her name permanently in the annals of show business. (Gilda's husband, Gil Boag, was co-owner of the Folies Bergère, a coincidence which in no way obstructed her future.)

On opening night Gilda tripped out to the center of the dance floor in a bright red dress glittering with spangles and, while the Dixieland Band whipped up a chorus of "Indianola," she generated a chassis movement that singed every tablecloth within six paces. A slightly inebriated customer at a nearby table shouted, "Hey, what do you call that?"

"I'm shaking my shimmy, that's what I'm doing," chirped Gilda.

Her stay at the Folies Bergère was quite brief. In a letter to Emile Christian (now back in London) only nine days after opening night, LaRocca wrote as follows:

. . . Well, boy, we're cleaning them up around here. Our come-back is great and shows Mitchell was wrong when he said they would not stand for our music on Broadway any more. Well, the girl who was working with us was let go, as the people just went wild over our return and didn't need any other attraction . . . When we get working together in about three more months we will have a band that will run anything off Broadway. As we stand, we clean them out. . .

The golden-haired singer went on to shake her shimmy in the Ziegfeld Follies and was shortly the rage of the nation. As her fame began to spread she sought to fan the fire by buying double-page spreads in *Variety,* the very publication that had only a year earlier derided shimmy as "a vulgar cooch dance at best."

But the Dixieland Band suffered no shortage of shimmy within its own ranks. Nick LaRocca, whose dancing talents rivaled his musical genius, often stepped down from the bandstand to dance with the customers or "shake a wicked shoulder" in his own version of the shimmy. *Variety,* in describing the Dixieland Band in vaudeville, remarked with restrained surprise that "one of the players shimmies while playing." Regular patrons at the Folies frequently requested this dance routine, the movements of which were improvised in a manner paralleling the music to which it was danced. Another left-handed cornetist, Sharkey Bonano, who is no slouch himself at the impromptu shimmy, still remarks with enthusiasm, "Man, could that boy dance!"

The popularization of the shimmy brought about a wave of shimmy songs, such as "Shim-Me-Sha-Wab-ble," "Shake It and Break It," and "I Wish I Could Shimmy Like My Sister Kate." But oddly enough, these tunes were not in a suitable rhythm for shimmy and were usually sung rather than danced. The Original Dixieland Jazz Band was responsible for the popularity of "Sister Kate" in the North long before the song was published as sheet music in 1919; and its New Orleans composer, A. J. Piron, entered into correspondence with Nick LaRocca relative to its recording. In a letter dated August 13, 1922, he commented as follows:

My Dear Mr. LaRocca:
. . . In reference to paragraph 5 of your letter, will say that I am unable to make a complete statement concerning turning it over to a large publishing concern, as I have already several contracts from different concerns to be considered, but in the meantime I will appreciate anything that you might do to record "Sister Kate" or any other number of mine, and you can rest assured that you will be greatly recompensated for same . . .
  For some time I have felt that the title "I Wish I Could Shimmy Like My Sister Kate" was a little out of date, and for that reason I have changed it to "Sister Kate" with a rearrangement of lyrics, and in a few days you will receive the new outfit which I hope you can use to a great advantage.
  Now, trusting that you will be able to use the numbers I am sending under separate cover, I am

Yours very truly,
A. J. Piron

It is unfortunate that "Sister Kate" was never recorded by the Dixieland Band, and it is logical to assume that the financial arrangements were not made attractive enough to LaRocca.

But the most popular new hit of the twenties came from the poetic imagination of the man who sat at the keyboard of the Original Dixieland Jazz Band. One afternoon at the Folies Bergère, J. Russel Robinson arrived at rehearsal with the lead sheet of his latest brain child, a ditty he called simply "Margie." He picked out the melody on the piano, as LaRocca and Shields stared over his shoulder at the music which was to them only meaningless spots on lined paper. Edwards, with a mute stuck in his trombone, quietly worked out the harmony he intended to play when the new song was given a try that evening.

Needless to say, "Margie" was an instant success. The Dixieland Band played the tune twice nightly at the Folies, LaRocca singing the words to enthusiastic patrons who stopped dancing to listen. Within the first week both Eddie Cantor and Al Jolson were hypnotizing their audiences with the first "girl" song since "Jenny."

It was "Margie" that heralded the new series of Victor phonograph records by the Dixieland Band. On November 24, 1920, the musicians gathered at the recording studios to pick up where they had left off a few years earlier. If they had had their own way, they might have waxed six more jazz masterpieces. But E. T. King and J. S. MacDonald of the Victor Company had other ideas. They no longer wanted the lively, uninhibited playing that had ignited the flame of the new musical age in 1917, but wanted instead a new kind of "sweet" jazz. King had been at the Folies a few nights before, had heard Robinson's new hit, and now insisted that "Margie" be the first recording of the new series.

The musicians, even including composer Robinson, protested hotly, but the decision had been made. And if this were not enough, Messrs. King and MacDonald also demanded that the band keep up-to-date by adding a saxophone. The records that were cut during 1920 and 1921 by this augmented group demonstrate how the addition of another wind instrument destroys the classic pattern of five-piece dixieland jazz. The tenor sax conflicts with the trumpet, having nearly the same range, and is only effective in specially arranged passages. For the most part the trumpet (often muted) plays in unison with the saxophone; the saxophone, probably closer to the pickup horn, dominates the melody. There are sections of these records where the saxophone is used with interesting results, but they do not make up for the deadening effect on other parts of the ensemble. If too many cooks can spoil the broth, this is one indigestive example. Even as played by the accomplished Benny Krueger—undoubtedly the greatest sax technician of that period—the instrument fails to find a niche for itself among the complex interlacing strains of the music.

LaRocca, too, had gone "modern" by substituting a trumpet for his faithful cornet, the latter having been patched and welded in two places where the bandleader's fingers had worn completely through the brass! If any musical instrument were ever a monument to the ardor of its master, the original cornet of Nick LaRocca holds that distinction.

The relunctance of old-time musicians to switch instruments or buy new ones is best explained by Eddie Edwards. "Musical instruments are like automobiles," he says. "Nowadays almost any car you buy is a good one, but in the old days if you found a good one you considered yourself lucky. It was the same way with musical instruments, especially horns. When you finally located a good one, nobody could make you part with it."

The commercial influences of the early twenties are

readily apparent in the six Victor records released by the Dixieland Band during 1920 and 1921 (see Table 7). Conspicuously absent is the uninhibited, freely creative spirit of the old Dixieland five. Gone is that close, sensitive co-operation which made the first six records an historic work of genius. LaRocca still carries the melody as he alone can, but this time it is someone else's, for the Original Dixieland compositions have given way to the catchy but ephemeral songs of Tin Pan Alley. The fox trot has now replaced the one-step entirely, as well as nearly all other forms. Shields still "noodles" on his clarinet, but only when he feels he can get away with it.

Nevertheless, the experienced judgement of King and MacDonald proved correct. After all, the Victor Company was a profit-making organization interested primarily in sales. In this respect, the end product was more than gratifying. "Margie" flooded the market when it was first released in 1920, ringing up sales far into the hundreds of thousands of copies. Today it is the most common recorded example of the Original Dixieland Jazz Band to be found anywhere in the world, even if artistically by far the worst.

It was originally decided to back up "Margie" with the Dixieland Band's distinctive arrangement of "In the Dusk." But although two master records of this number were cut—one on November 24 and the other on December 1—apparently the Victor people felt that too much of the old hot jazz was still showing through. "In the Dusk" was accordingly junked and a more melodic substitute called for. Here LaRocca stepped in and once

TABLE 7

Recordings of the Original Dixieland Jazz Band
(1920–1923 Series)

RECORDING DATE	COMPANY	NUMBER	TITLE	RELEASE DATE
Nov. 24 1920	Victor	——	Margie	Not Released
		——	In the Dusk	Not Released
Dec. 1 1920	Victor	——	In the Dusk	Not Released
		——	Satanic Blues	Not Released
		——	'Lasses Candy	Not Released
Dec. 4 1920	Victor	18717	Margie	*ca.* Dec. 1920
			Palesteena	
Dec. 30 1920	Victor	18722	Broadway Rose	*ca.* Jan. 1921
			Sweet Mamma	
Jan. 28 1921	Victor	18729	Home Again Blues	*ca.* Feb. 1921
			Crazy Blues	
May 3 1921	Victor	——	St. Louis Blues	Not Released
May 3 1921	Victor	18772	Jazz Me Blues	*ca.* June 1921
May 25 1921			St. Louis Blues	
May 25 1921	Victor	——	Satanic Blues	Not Released
May 25 1921	Victor	18798	Royal Garden Blues	*ca.* July 1921
June 7 1921			Dangerous Blues	
Dec. 1 1921	Victor	18850	Bow Wow Blues	*ca.* Dec. 1921
Jan. 2 1923	Okeh	4738	Toddlin' Blues	March 1, 1923
			Some of These Days	
Jan. 2 1923	Okeh	4841	Tiger Rag	March 1, 1923
			Barnyard Blues	

TABLE 7—*Continued*

Personnel (Victor Records):

Trumpet:	Nick LaRocca	Saxophone:	Benny Krueger
Trombone:	Eddie Edwards	Drums:	Tony Sbarbaro
Clarinet:	Larry Shields	Piano (first 11 sides):	J. Russel Robinson
		(last 7 sides):	Frank Signorelli

Personnel (OKeh) Records:

Trumpet:	Nick LaRocca	Saxophone:	Don Parker
Trombone:	Eddie Edwards	Drums:	Tony Sbarbaro
Clarinet:	Artie Seaberg	Piano:	Henry Vanicelli

more fought valiantly to have more of the Dixieland Band's popular standards recorded. There appears to have been an attempt at this on that same day, when both "Satanic Blues" and "'Lasses Candy" were again put on wax. But, just as they had done in 1918, Mc-Donald and King again turned thumbs down on these two LaRocca originals. As far as they were concerned, if you couldn't sing them, you couldn't sell them. Not only that, but they weren't yet satisfied with the recording of "Margie." So the band was called back to the Victor studios the following Saturday, December 4, when they cut their second platter of "Margie," adding Robinson's "Singin' the Blues" to form a medley, and grooved the reverse side with still another of Robinson's hits then being sung up and down Broadway, the exotic "Palesteena" ("Lena is the Queen of Palesteena"). Hence this top-selling record, graced by three of his original compositions, marked J. Russel Robinson's shining hour in musical history.

The "Margie"/"Palesteena" combination was issued in England three months later under the HMV ("His Master's Voice") label by Victor's British affili-

ate, the Gramophone Company, Ltd., and it nearly involved the Dixieland Band in another lawsuit. The
Columbia Graphophone Company claimed that the release violated its contract with the Dixieland Band,
which provided that the New Orleans musicians were not
to "make any records whatever for reproduction by any
sound recording or sound reproducing machine or device for any talking machine or record manufacturer in
the United Kingdom save for Columbia." As a conse-
quence thereof, Columbia advised they were discontinu-
ing payment of royalties to the band. LaRocca, now
acting as his own attorney, pointed out that the second
Victor series was made after the band had left the ter-
ritorial limits of the United Kingdom and that an artist,
by the very nature of his work, is unable to control the
sales of his product. Through an involved correspondence he finally succeeded in recovering the royalties,
which evidently were not considered worth fighting over
by the company, the HMV records of the Dixieland Band
having by this time reduced sales of the Columbia records to a negligible amount. Thus the band had suffered
the strange fate of competing with itself!

"Broadway Rose" was recorded thirty days later,
as the band attempted to duplicate the smashing success
of "Margie." A daring experiment is tried in the title
song, with trumpet, trombone, and saxophone playing
melody in unison. After the medley tune the trombone
again returns to the foreground to take a straight solo
on the second chorus, with Benny Krueger showing off
his well-developed technique and the other muted instruments forming a background. The "home stretch" is
more conventional dixieland, and J. Russel Robinson

does some pleasing work if the listener is able to catch it.

The verse and chorus of "Sweet Mama," on the flip side, are followed with a chorus of "Struttin' Lizzy," featuring a clarinet-saxophone duet in which the clarinet improvises about the saxophone melody and takes a well-executed break. The musicians do something vaguely resembling singing in certain sections of this record; at the end Nick LaRocca, from his position twenty feet from the pickup horn, drops his trumpet to shout, "Yessuh! Sweet Mama! Papa's gettin' mad!" This is the only Dixieland Band recording in which the voice of a member is heard. (This ending phrase was often used by LaRocca at dances to whip up audience reaction. After "Tiger Rag," for example, he would be heard to yell, "Yes-suh! That's the Tiger!")

"Home Again Blues" is easily the best of the series. The first part undoubtedly contains the most powerful, confident, sharply cut trombone breaks ever recorded. If they sound a bit "corny" to the modern musician, he may well note the date of the record, for the originators of a "hot lick" must never be blamed for the number of times it is repeated. The second chorus illustrates La-Rocca's newly acquired practice of playing counterpoint and contains an outstanding break by Larry Shields. "Crazy Blues," on the other side, "travels" in a steady, unbroken course. After an interesting rendition of "It's Right Here For You," the trombone steps to the spotlight in a solo of deep, rich tones, as Krueger improvises about the melody.

Although most of these renditions lack the sincerity of the first enthusiastic efforts of the band in 1917, occasional snatches of really beautiful jazz are

still to be heard, isolated areas where the fervid spirit
of the old five breaks through the hard, glossy finish of
the first commercial jazz. Shields and Krueger work
very well together. Their duet in the second half of
"Jazz Me Blues," a device wherein the clarinet rides
over the saxophone melody, is executed with great feeling.

After recording "Jazz Me Blues," three weeks
elapsed before a decision was reached on a companion-
piece. During this interval LaRocca was approached by
W. C. Handy and his brother, partners in a music publishing firm known as Pace and Handy. Their business
was tottering and they were hoping for one more song
hit to put them back on their feet. LaRocca, with many
of his own compositions yet unrecorded, was not looking
for new material. He finally agreed to plug Handy's
"St. Louis Blues," a song originally published in 1914,
in return for a 10 per cent commission on music sales.
The recording of "St. Louis Blues" sold more than
300,000 copies, sheet music sales soared, and the Handy
brothers were back in business. LaRocca, however, was
paid off with a rubber check which now occupies a space
in his scrapbook.

The Original Dixieland version of "St. Louis
Blues" is distinguished by its famous clarinet solo,
orginated by Larry Shields and imitated so many times
in the succeeding decade that it became a jazz classic.

J. S. MacDonald, not yet content with his stream-
lining of the Dixieland Band, now decided that the band
needed a vocalist. In New Orleans he discovered Al
Bernard, a blackface comedian and minstrel show
singer, and brought him north for the job. Bernard

vainly attempts to make himself heard on "Royal Garden Blues" and "Dangerous Blues," but it appears that the band is fighting him to the finish. Frank Signorelli, who has by now replaced J. Russel Robinson on piano and appears a bit overzealous in his accompaniment, doesn't really give the New Orleans boy much of a chance. Perhaps it is just as well, for it gives us an opportunity to hear some fine bang-out pianoforte by this inspired musician. The background turns out to be far more interesting than the featured artist.

Not to be overlooked by any means is LaRocca's torrid trumpet break in "Dangerous Blues," as this one easily ranks with his classic "flying tackle" breaks in "Fidgety Feet" and "At the Jazz Band Ball" on the old Victors. It appears that Nick, who had been subdued, restrained, muted, or possibly just discouraged in the first few records of this series, gradually moved forward as further recordings were made, steadily bringing the band back to life as he did so. Those ripping phrases, the tone that cuts like a knife, the powerful drive that no one has ever imitated made the last few platters of this series an atonement for the first. It is further proof that LaRocca himself was the heart of Original Dixieland Jazz Band, and that as LaRocca went, so went the band.

"Bow Wow Blues," a slow, doleful number characterized by a barking dog that monopolizes all the breaks (ranging from the falsetto "yipe" of a poodle to the basso profundo "ruff" of a hound) and makes himself heard again in the two-beat rest preceding the dixieland tag, is the last of the series. This record, with its opposite side grooved by the Benson orchestra of

Chicago, sold very few copies and is undoubtedly the rarest item of both series. The Original Dixieland Jazz Band, once the world's greatest recording combination, bowed out with the "Bow Wow Blues" and didn't return to Victor for fifteen years.

But the reactionary antijazz forces that had begun to make themselves felt in the recording industry had not yet reached Broadway. The old five-piece combination, minus saxophone and vocalist, continued to bring down the house at the Folies Bergère Café without any supporting act. On January 24, 1921, an enthusiastic management renewed their contract for another twenty weeks.

Then, just as the band had begun to pick up steam, more personnel problems—more of those squabbles that often plagued the Dixieland Band and were ultimately responsible for its collapse—resulted in new changes. An unfortunate misunderstanding had taken place with regard to Robinson's smash hit "Margie," an argument ensued between the composer and other members of the band, there was the inevitable flare-up of temperaments, and LaRocca gave Robinson his two weeks' notice. LaRocca claims that Robinson had defaulted on a promise to pay the band a bonus in the event "Margie" went over.

He was replaced on April 11 by twenty-year-old Frank Signorelli, a pianist who was already filling up the keyboard with more fingers than anyone else in the business. Born in New York City on May 24, 1901, he had taken piano lessons only two years before being overcome by a strong compulsion to earn a living at the ivories. At the age of seventeen he was playing his first

professional job at an East Side café called the Central
Opera House, accompanying a popular blues singer
named Tess Giardella who appeared nightly in blackface, billed as "Aunt Jemima." It was here during
1918 that Nick LaRocca and Tony Sbarbaro frequently
dropped by on their off-hours from Reisenweber's to
hear Frankie attack the keyboard with an avalanche of
real ragtime. LaRocca and Signorelli later became close
friends and often went to parties together, Frankie playing the piano, and Nick, with seemingly inexhaustible
energy, putting on his own song and dance act.

With Signorelli on piano the Original Dixieland
Jazz Band finished out another successful two months at
the Folies, unaware of the long chain of misfortunes
which was about to overtake them.

13

Polar Bears and Red Hot Mammas

Man from both sides of the great dividing ocean had now expressed his views, more often than not in pitiful confusion, on the new jazz music. But what of the other mammalian species? What was the unspoken opinion of the savage beast himself? Was he really soothed by the charms of music? A certain Dr. Norman Spier, who introduced himself to the scientific world as "a psychologist at large" (but neglecting to say from what), decided to have this problem settled once and for all.

From behind the wrought iron gates of New York's Central Park Zoo there came, on the afternoon of April 20, 1921, sounds totally foreign to the animal world. Strollers through the park that day may have been startled to hear, high above the hysterical screams of the hyena, the frenzied wail of a jazz clarinet, while the brassy blare of a cornet mixed with the chattering of orangutans. Further wafts of the spring breeze may have brought forth a strange and undignified conversation between the mighty lion and the snarling masculine voice of a tailgate trombone.

The more musical of these varied sounds emanated from the horns of the Original Dixieland Jazz Band, which was that day giving a command performance of "Barnyard Blues" for the benefit of scientists and savage beasts arranged on their respective sides of the cages in the Central Park menagerie.

Scientists who had accepted this unusual invitation of Dr. Spier included Dr. E. H. Pike and Professor R. I. Scott of the Physiology Department of Columbia University, Dr. H. D. Jones of the same faculty, and a collection of zoologists from New York's Museum of Natural History. Park Commissioner Francis Gallatin had granted Dr. Spier special permission for the experiment, which was conducted under the cold and penetrating gaze of Dr. Reid Blain of the New York Zoological Park.

For press agents, it was a dream à la mode. For newspaper reporters, who came there in droves, it was the opportunity of a lifetime. They ran through unexplored metaphors in their bare feet, and no figure of speech lay untrampled. The inevitable crowd of onlookers gathered but, in order not to excite the animals, only interested officials, scientists, press agents, journalists, photographers, and newsreel cameramen were admitted.

The first ensemble number, "Sweet Mamma," was played in the Lion House for the benefit of Miss Murphy, a huge hippopotamus. Miss Murphy had been a member of the New York zoo family for thirty years and was the mother of fourteen baby hippopotami, now scattered over all parts of the world, but she undoubtedly hadn't heard anything like this in all her born days. According to the New York *Daily News*, Miss Murphy had been "leading an orderly life with her son Caleb, with never a complaint from the neighbors. When she heard jazz right at her front door, she went diving into the tank, taking the boy along, and she wouldn't come up until the musicians moved along."

"Ostrich Walk" was played for the first time before the lion family. Ready to express a frank opinion were Helen the lioness, and her three offspring, Edmes, Cleo, and the giant male Ackbar. Edmes, as observed by the New York *Morning Telegraph*, "pleased the musicians by coming in with one note on the first break of the tune. But Ackbar gave every evidence of dislike, rearing furiously and lashing his tail."

The jazzmen moved on to the cage of Bagheeta, the black leopard, and struck up a chorus of "Satanic Blues," followed by several more lively jazz numbers. As described by the New York *World*, "he had lain like a black coal in a dark corner, two burning points where his eyes were, and a red, hissing mouth, when the jazz was played. But at 'Lead, Kindly Light,' he flew at the bars, tearing toward freedom and the musicians, a picture of rage."

"It was fifty-fifty with Bagheeta," adds the New York *Daily News*. "He couldn't fall for jazz, but he assumed a front row nonchalance just to show he was a good fellow, until they tried 'Lead, Kindly Light' on him, and then he jumped for the players. Bagheeta would stand the jazz, but no hymns went in his young life."

Willis Holly, secretary of the Park Department and obviously no jazz fan, made the cynical remark, "I don't see how this experiment will determine anything of scientific value until the band begins to play some real music. So far I haven't heard any."

Commissioner Gallatin replied, "Probably the effect would be different if we should put the musicians inside the cage."

A lithe and graceful puma exhibited the nearest approach to appreciation shown by any animal in the building. When Eddie Edwards picked up his violin to play "Maid of the Mountains," a dreamy waltz, she stepped across the floor and made half a dozen circles of her cage in time with the music.

In the interests of science and a photographer, Mlle. Evelyn Valee, a beautiful Parisian dancer, was holding Billy Buck, a six-months-old kid, in her arms when the five-piece combination of the Original Dixieland Band suddenly cut loose. Billy Buck did likewise, with spectacular results. "He wriggled end for end in the arms of Mlle. Valee," reports the *Tribune*, "and the beautiful Parisian danseuse, presenting a violently protesting tail to the Dixieland Jazz Band, was butted right in her beautiful Parisian frock. Mlle. Valee said 'umph' in beautiful Parisian and slung Billy Buck with beautiful precision right at the saxophone player, from whom he rebounded and scuttled away."

Betsy and Jewell, the elephants, seemed to enjoy the music immensely, but Dr. Pike was unimpressed. "These elephants have at some time been on the road with a circus," he declared. "They have a sort of circus instinct and welcome any form of entertainment."

Ringtailed monkeys shook their cage doors and screamed, chattered, and raved, some with obvious delight and some with wrath. Big blue-nosed apes did jungle war dances; Joe the chimp wept bitterly. Others swung by their hands from a trapeze and banged their feet rapidly against the wire netting of the cage. The wilder the music, the better they liked it. "Ostrich Walk" threw most of the monkeys into shimmy dancing, which caused bandleader LaRocca to recall that music sometimes made human beings dance like monkeys.

But the star of the afternoon was Jim the polar bear. Although driven back to his cave under the rocks when the full band played, he came out and shimmied gracefully when, at the suggestion of Dr. Pike, Shields and Edwards harmonized soothingly on clarinet and trombone. The New York *Evening Journal* states that "of all animals, Jim seemed the most appreciative. He actually did a shimmy, starting the movement at his nose as he cavorted at the top of his cliff cave and winding up with a genuine demonstration of the terpsichorean art all over his body." While the learned Dr. Pike offered an involved scientific explanation of the strange shiver, Head Keeper James Coyle merely remarked: "My Boy, I never seen that bear act that-a-way before!"

On June 11 the Original Dixieland Jazz Band closed at the Folies Bergère and opened four days later at Café LaMarne, on the boardwalk at Atlantic City. For the twelve-week run they were paid eight hundred dollars per week, plus 50 per cent of the cover charges over three hundred dollars. Alternating on the same program was Sophie Tucker, "The last of the Red Hot Mammas," who appeared with her own orchestra, billed as the Five Kings of Syncopation. Sophie had started as a blackface comedienne in vaudeville in 1906. After hearing a small-time theatrical manager remark that she was "too big and ugly" to appear on the stage without a disguise, she had gone on to fame in the Ziegfeld Follies and had recently scored further successes in her

Broadway show, *Hello Alexander*. She had by now abandoned the "coon shouting" style of singing she had used in vaudeville for the controversial *double-entendre* songs that were now shocking the more sensitive of night club patrons. At the LaMarne she sang such dubious numbers as "He Took it Away From Me" and "It's All Over," changing the lyrics only slightly to achieve an entirely different meaning.

Sophie was an inveterate card player and spent many of her off-stage moments in poker games with members of her orchestra, usually taking them for all they were worth. Edwards recalls that Larry Shields, who had an exceptional mind for mathematics, often sat in with Miss Tucker at these games and did more than hold his own.

In the fall of that year, the band went on a tour of Pennsylvania, playing a string of one-nighters that proved far more lucrative than steady engagements. Even small towns, hungry for "name" artists and willing to pay the price for a single evening, shelled out as much as $2,400 per night to engage the Original Dixieland Jazz Band, advertised widely as "The Creators of Jazz," for a local dance. Ticket prices were sometimes high, but farmers and coal miners drove to town over hundreds of miles to jam the dance halls on these occasions.

In Altoona, Pennsylvania, on Tuesday, November 8, 1921, Fred Oeffinger and C. J. Shimminger booked the band for a special dance at the Wolf Building. In order to prove that this was "The World's Greatest Jazz Band direct from the LaMarne, Atlantic City" as proclaimed on the billboard, they inserted the following announcement in their printed dance program:

TEST FOR ORIGINALITY

Two or more of the Original Dixieland Jazz Band recordings will be rendered on the Victrola sometime during the Dance. Immediately following, the orchestra in person will play the same selections, proving that you are getting the Original Band.

The highest paid band ever engaged to play for dance in this city.

The committee guarantees this to be the original Dixieland Jazz Band or will refund money.

In many places special tickets were offered at reduced rates for those who just came to stand and listen. But regardless of the publicity value of the world's first jazz band, which could now be advertised as "The Sensation of Two Continents," not all booking agents took these prices lying down. Phil Harris, manager of the Dance Guild of Scranton, Pennsylvania ("Bookers of High Class Dance Orchestras"), wrote LaRocca:

Also wish to let you know that we have had Ted Lewis' and Paul Whiteman's bands but at a much lower figure, and think it will be very hard for you to book at a price solid to any dance promoter in this state.

If you could consider a proposition of $2000.00 a week, I might be able to use the band. We have a hall in this town that holds 3000 people and would run your band at $500.00 guarantee on a 50/50 basis.

Kindly do not book up your band for this town until you hear from me in the next three or four days . . .

The manager of Charlton Hall, in Pottsville, Pennsylvania, "The Most Beautiful Ballroom in the Anthracite," penned a similar response:

I can readily see that you never played dance work, when you make me a price that is impossible . . .

I book for twenty-seven promoters throughout this section of the state. If I were to ask them for $1250.00 for one night, they would think I were crazy. I have handled every band in the country but your band . . . I am playing Whiteman himself and nine men, and they are getting just half what you ask.

But for every booking agent that turned them down, two would accept, and the Dixieland Band continued to earn more money with five men than Whiteman did with ten.

The Original Dixieland Jazz Band might have carried its successful tour into every state in the union, had it not been for the replacement problems that continued to haunt it. The temperamental Larry Shields, who had threatened to quit so many times that his warning was no longer taken seriously, submitted his two-week notice to LaRocca on November 29, just seven days before the band was to begin its two-week run at Jack Fiegle's Dance Palace in Philadelphia. Larry always hated New York, which he called a "cheese town," and now the endless town-hopping and living out of a suitcase had worn him down. Homesick for his wife, who ran a beauty parlor in Hollywood, he was now determined to settle down in California. He remained with the band for the duration of the Jack Fiegle job, and on December 20, 1921, unscrewed his clarinet and packed it neatly into its long, narrow box.

Thus ended the career of the colorful Larry Shields, considered by some to be the greatest jazz clarinetist of all time. Some historians have, in fact, mentioned Shields to the exclusion of any other musician in the Dixieland Band. This is an unfair exaggeration and is undoubtedly the result of desultory and superficial analysis of the band's famous Victor records, on which the clarinet of Shields cuts through the ensemble with extreme brilliance, owing to the sensitive high-frequency response of the old mechanical recording equipment. It was almost as if the recording device had been designed especially for Larry's clarinet. Nevertheless, Larry Shields will go down in history as the father of the "noodling" style and possessor of one of the most powerful clarinet tones on record. Heard in person, his volume was nothing short of astonishing. He used a very hard reed, and Edwards recalls that Larry would pick up and use reeds that other clarinetists had discarded as useless. Some of his runs have never been exactly duplicated, although many have tried, and some experts have attributed this to the fact that Larry—a self-taught musician—fingered his instrument incorrectly.

Finding a job in California was, at the offset, no great problem for Shields. Among the vast army of fans he had won at Reisenweber's before the war were George Jessel, Lou Holtz, Fannie Brice, Lou Clayton, Sammy White, and a hundred others now successful in Hollywood. They had tempted him with jobs in the magic Western city and were quick to make good their promises. Through his many friends Larry landed job after job in the film capital and elsewhere in California.

Six years later, with the advent of "talking pictures," he was frequently hired for bit parts in the movies, appearing as clarinetist in café scenes. But as his friends disappeared from the limelight and lost their influence, the jobs became fewer and farther between, and Larry's career as an actor was short-lived.

The loss of Larry Shields was a serious one for the Original Dixieland Jazz Band. On December 24, 1921, LaRocca returned to New York to sign the first of a long series of contracts with Sixti Busoni, owner of dance halls in New York, Brooklyn, and Coney Island. At the conclusion of the job at Jack Fiegle's Dance Palace in Philadelphia, the band moved to Busoni's Balconades Ballroom at Sixty-sixth Street and Columbus Avenue to begin a twenty-week engagement with Johnny Costello on clarinet. Costello had been a member of the first Memphis Five combination that had been formed at Coney Island in 1919, which had included Phil Napoleon and Frank Signorelli.

The Dixieland Band was now back on Broadway, even though its salary had dropped from a thousand dollars a week to eight hundred and fifty. Edwards complained of this cut and felt that LaRocca was selling the band short, but the fact remains that the band was kept in business and was therefore able to maintain its popularity under increasingly adverse conditions.

At the Balconades, a personal feud between Johnny Costello and Frank Signorelli quickly came to a head. When a choice had to be made, Costello was the one to leave and was replaced by fourteen-year-old James Sarrapede, later to become known to the music world as Jimmy Lytel. The student of jazz history may now note the nucleus of the Original Memphis Five shaping up within the ranks of the Dixieland Band. It was now only necessary for one of those peculiar machinations of fate to do the rest. On March 15, 1922, Nick LaRocca re-

turned to his hotel room with an advanced case of pneumonia, and the Balconades soon found itself with a ballroom full of dancers and no jazz band. Quick to fill the
vacancy, Frank Signorelli left LaRocca to form the
Original Memphis Five, taking along clarinetist Jimmy
Lytel. This new band, which was to become second only
to the Original Dixieland in historical significance, also
included Phil Napoleon on trumpet, Miff Mole on trombone, and Jack Roth on drums. Phil and Miff had been
frequent visitors to the Balconades.

However, the relationship of these two great organizations neither began nor ended at this point in history.
The first band to call itself the Memphis Five—Phil
Napoleon (cornet), Johnny Costello (clarinet), Moe
Gappell (trombone), Frank Signorelli (piano), and
"Sticks" Kronengold (drums)—had been formed at
Coney Island while the Dixieland Band was in England,
although Signorelli is quick to point out that this band
"really didn't amount to much" and seems reluctant to
give it mention. Its members were still adolescents during the Coney Island phase and it is reasonable to assume that they were still suffering the musical growing
pains so characteristic of the traditional, self-taught
jazzman. For if there is one distinguishing element in
the makeup of an old-time jazz musician, it is that he
learned the hard way and often fought an up-hill battle
to master his instrument. It was frequently from this
struggle that an original style was born.

Phil Napoleon had been the advance guard of that
growing group of young musicians that had haunted
Reisenweber's almost every night before the war when
jazz was beginning to take its grip on New York. One of

the most rabid fans of the new music, the young Napoleon called frequently at LaRocca's hotel room to find out how certain "licks" were executed, or to gain other musical pointers from the popular bandleader.

When the Dixieland Band returned from England a few years later, Napoleon was surrounded by a whole cult of up-and-coming jazz musicians. These enthusiastic youngsters, who were much in demand in jazz-hungry New York, often gathered with members of the Dixieland Band "after hours" for jam sessions and the inevitable rounds of nocturnal revelry, in which girl friends played no little part. This colorful weld of New York and New Orleans jazzmen painted the town a gaudy vermilion on more than one occasion, witness the following whimsical item in the New York *Daily News* of May 15, 1921, while the Dixieland Band was still employed at the Folies Bergère:

25 SEIZED IN RAID BY COPS, WHO POSE AS PARTY'S GUESTS

Ten Girls and Fifteen Men "Too Noisy"
Celebrants of Jazz Band's Return

SPILL ALLEGED HOOCH

Dry Sleuths Go Through Reisenweber's
Wall and Raid Circle Hotel

Ten young women and fifteen men, all actresses, actors or musicians, were locked up in West 30th Street Station early to-day on charges of disorderly conduct, after Detectives Levine and Sheehan of Inspector Boettler's staff had broken up what the prisoners indignantly declared was a respectable private party.

The detectives said they were passing No. 111 West 49th

Street at 3 A.M. when they heard a jazz band and other distressing sounds from a lighted room on the top floor. They thought it was too much noise for that time of night and went up. The man who opened the door was not cordial until the detectives said they were from Greenwich Village and would like to join the party.

They were admitted to a small room in which men and women were sitting on beds, boxes and the floor. The band was playing. When all raised their glasses to drink a toast, the detectives showed their shields, whereupon every glass crashed to the floor and the contents spilled.

After telling the men and women they were under arrest, the detectives snooped around until they found a suitcase containing several bottles of supposed hooch, but they could find no one who claimed ownership, so they charged their prisoners with disorderly conduct and took them to West 47th Street Station in two patrol wagons.

The prisoners declared there was nothing improper about the party. They said it was in honor of the first anniversary of the jazz band's return from Europe—it is a band well known on Broadway—and that it was given late because they were all professional people and many did not finish work until after midnight. Their resentment at having the party spoiled and being carted to the station in the patrol was nothing to what it was when they learned they would have to be locked up for the rest of the night.

The accompanying photograph showed a line of musicians and actors, coats over their arms, prominently including Napoleon, Signorelli, Sbarbaro, and Shields. That wise old team of LaRocca and Edwards had evaded the press photographer by ducking behind a desk before the flash went off. Although the prisoners were fined only a dollar each, LaRocca complains that the judge confiscated his phonograph records and never returned them. All of which goes to show that there's a jazz fan in every precinct.

The Balconades job launched the Original Memphis Five on a long and spectacular career. Although active for years in Manhattan, they perhaps achieved their greatest fame via the phonograph. Willing to record for less money than the Original Dixieland, they turned out a prodigious amount of very good jazz on such labels as Brunswick, Vocalion, Regal, Cameo, Gennett, Pathé, Perfect, Grey Gull, Paramount, Puritan, Buddy, Lincoln, Triangle, Banner, Imperial, and Melotone, as well as a respectable assortment for the two leaders in the field, Victor and Columbia. For this army of companies they cut no less than 225 sides between 1922 and 1927! Although nearly all bear the name of the Original Memphis Five, a few of these were recorded under the names of Ladd's Black Aces, the Cotton Pickers, and Jazz Bo's Carolina Serenaders. Charles DeLaunay's *Hot Discography* carries a complete listing.

Among the better sellers, the earliest records are perhaps the most interesting. "Pickles" (Columbia A 3924), recorded in 1924, offers a good example of Frank Signorelli's piano work by featuring him for a full half chorus without so much as the benefit of drum accompaniment, but with the kind of bass that Signorelli pounded out certainly no drums were needed. "Aggravatin' Papa" (Vocalion A14506) and "Snakes Hips" (Victor 19052), cut in 1922 and 1923, reveal the intense trumpet drive of Phil Napoleon. Napoleon carried with him some of LaRocca's innovations—the trick of leading into a chorus with a "dirty" note, or dropping out on the first two beats of a leading measure—but he also added many of his own rhythmic ideas.

The Original Memphis Five specialized in tricky,

intricate arrangements but, according to Signorelli, never played their introductions from written scores—as has been claimed by certain writers. They exploited the principle of the solo "break" to its fullest, just as the Dixieland had done, and those contributed by Napoleon, Mole, and Lytel are works of art. They recorded "I Wish I Could Shimmy Like My Sister Kate" six times on six different labels, but Nick LaRocca contends that their arrangement is the one used by the Original Dixieland Jazz Band when Frank Signorelli was playing with them at the Balconades.

The life-span of the Memphis Five, under its original leadership, was no more than a brief ten years, but during that time it fostered some of the greatest names in American dance music. Tommy Dorsey, Jimmy Dorsey, Red Nichols, Frankie Trumbauer, and Eddie Lang all had apprenticeship in this organization.

The merits of Signorelli's old band in the field of what is now called "dixieland jazz" have been debated for decades, but of one thing we can be sure. For a bunch of boys that had never been any farther south than Coney Island, the Original Memphis Five did all right for themselves!

14

The March of the Moralists

The antijazz movement that had started during the early postwar years now gathered momentum as indignant parents, educators, clergymen, and opportunist politicians sought to crush the music that had become the symbol of a wild new era. These horrified members of an older generation had tolerated jazz with disgust or mild amusement while it remained behind cabaret doors, but now that the phonograph had brought it into millions of American homes and the dance halls of every large city resounded with wailing saxophones and clattering trap drums, they were moved to action.

It was not the music itself, of course, so much as the manners and customs associated with it that alarmed civic leaders. Shimmy dancing, bootleg hooch, female smoking, and premarital sex appeared as a shattering distortion of contemporary ethics. No tradition seemed too sacred. Even the heretofore stable English language seemed doomed, as the word "neck" rapidly changed from a noun to a verb. The elders saw it as the end of the world. This revolt of the younger generation was induced by deep-seated psychological ills, and while jazz may have been only the outward manifestation, it was everywhere recognized as the cause.

Dr. Henry Van Dyke, speaking before the National

Educational Association in Atlantic City on February 27, 1921, declared that "jazz music was invented by demons for the torture of imbeciles. The State has the same right to protect its citizens from deadly art as it has to prohibit the carrying of deadly weapons, but I do not think the law can reach the matter. It is spiritual. As teachers, let us not rely wholly on the law to make people virtuous . . ."

That same year Mrs. Anne Faulkner Oberndorfer, the national chairman of the Federation of Women's Clubs, advocated an all-out war on jazz. "We must familiarize ourselves with the music that is being used in our schools, clubs and homes," she declared. "We shall be surprised, even horrified, with what we shall find, but it is time we knew. Jazz in its original form was used as the accompaniment to voodoo ceremonies. Is it any wonder that the largest industries which started community singing during war times have been forced to forbid the singing of jazz in any of their factories?"

The New York *Herald* applauded this "movement to substitute music for the jingling clatter called jazz" and commented as follows:

It is an interesting plan of campaign. That it would at once exorcise the demons of jazz discord now rampant is probably too much to hope. Their name is legion. They have established a tenacious grip on the territory they have invaded. But with the Federation of Women's Clubs warring in earnest on jazz ultimate victory may be safely predicted. Jazz cannot stand the light of widespread education in real music, and the Women's Federated Clubs, under the leadership of Mrs. Oberndorfer, intends to turn a submerging flood of that light upon it . . .

As the antijazz crusade gathered momentum, the New York *American,* in its issue of January 22, 1922, expressed these dire warnings:

JAZZ RUINING GIRLS, DECLARES REFORMER

Degrading Music Even Common in "Society Circles,"
Says Vigilance Association Head

Chicago, Jan. 21—Moral disaster is coming to hundreds of young American girls through the pathological, nerve-irritating, sex-exciting music of jazz orchestras, according to the Illinois Vigilance Association.

In Chicago alone the association's representatives have traced the fall of 1,000 girls in the last two years to jazz music.

Girls in small towns, as well as the big cities, in poor homes and rich homes, are victims of the weird, insidious, neurotic music that accompanies modern dancing.

"The degrading music is common not only to disorderly places, but often to high school affairs, to expensive hotels and so-called society circles," declares Rev. Phillip Yarrow, superintendent of the Vigilance Association.

The report says that the vigilance society has no desire to abolish dancing, but seeks to awaken the public conscience to the present danger and future consequences of jazz music . . .

In Milwaukee, City Health Commissioner Dr. George C. Ruhland warned that "jazz music works up the nervous system until a veritable hysterical frenzy is reached. It's easy to see that such a frenzy is damaging to the nervous system and will undermine the health in no time. The shimmy and related dances should be avoided, not only because they are damaging, but they are unaesthetic and certainly not beautiful." He added that jazz was an "atavistic reversion to the primitive."

Cardinal Begin, the Archbishop of Quebec, condemned "lascivious" dances "in order to raise an effective dike against the rising flood of neo-paganism" and asserted that they must be combated as moral contagion. A decree of the Synod, published at the same time, named the prohibited dances:

> We energetically reprove those dances which are lascivious, either in themselves—such as the "fox-trot," the "tango," the "shimmy," the "cheek-to-cheek," the "turkey-trot," the camel-trot," the "one-step," "two-step," and others of the same kind, by whatever name they may be called—or in the manner in which they may be executed—as in the case with the waltz, the polka, and other dances which are commonly danced today in a lascivious manner . . .

The *Literary Digest* of March 22, 1924, echoed the public cry in a series of articles blasting the popular dances. In an article titled "Trotting to Perdition" it stated:

> Medical men are not noted as ethical extremists, and it is, therefore, all the more significant when they join the chorus of religious leaders and police matrons in condemning certain types of modern dances as relics of jungle days, saying, we are told, that these degenerate dances are as morally harmful among civilized peoples, and that they are to be regarded as traps set to ensnare innocent feet. In New York an amazing condition of immorality has been found to exist in twenty per cent of the public dance-halls . . .
>
> Immoral excesses of the worst sort exist in some of the dance-halls of New York, according to the report of a four month's survey . . . The extremely indelicate and immodest practices of these places constantly call for the most rigorous regulation by city authorities . . .

It was inevitable, then, that the indignation of the straight-laced element would worm itself into legislation. In April, 1922, New York State's Governor Nathan L. Miller signed the Cotillo Bill, which gave the Commissioner of Licenses in New York City full power to prohibit dancing wherever and whenever he saw fit. The New York *Herald* described the bill in these words:

> Dancing in all of Broadway's jazz palaces, as well as in the dingiest hall in the old Bowery district, is to be brought under municipal control . . . The Commissioner of Licenses in New York is to be dictator of dancing. He can decide whether the shimmy is immodest; just what kind of fox trot or turtle slide young girls may dance in public without endangering their morals and how late the youth and graybeard of the metropolis may dance at night without risking their health . . . These regulations are all contained in the Cotillo bill, which amends the city charter in relation to the regulation of dance halls.

The Commissioner, exercising his newly vested powers, thereupon outlawed all jazz and dancing on Broadway after midnight. This spelled the beginning of the end for jazz in New York City, as dance hall managers were now forced to lower musicians' wages to a level where such work was no longer profitable.

These were the conditions facing the Original Dixieland Jazz Band on April 10, 1922, when LaRocca, now recovered from pneumonia, reorganized his group for a two-week run at the Flatbush Theater in Brooklyn. Replacing Signorelli and Lytel, who were at the Balconades with the Original Memphis Five, were Artie Seaberg on clarinet and Henry Vanicelli on piano. The original nucleus of LaRocca, Edwards, and Sbarbaro still remained intact.

Following the Brooklyn job, Sixti Busoni sent them to his Rosebud Ballroom at Coney Island for the summer. They opened there on May 13 and moved over to Busoni's Coney Island Danceland two days after the Fourth of July.

While the band was engaged at Coney Island, LaRocca and Edwards lived within commuting distance at Sheepshead Bay, Long Island. Here they went into partnership on an eighteen-foot motorboat, which they used for fishing and short excursions up the Hudson to Poughkeepsie and the Pallisades. Fishing was almost a regular afternoon ritual for the two musicians, who worked all night and slept all morning. LaRocca recalls that they sometimes fished for drowned bodies to help the authorities and were even loaned a grappling iron by the police department. On stormy nights LaRocca (who could not swim) could be seen wading out into the Sound in water up to his neck to moor the boat, as waves broke over his head. Around his waist was a rope, the other end held on the shore by the laughing Edwards, ready to pull him in when the chore was completed.

LaRocca, who believes he was the first man to catch lobsters in Sheepshead Bay, constructed his own traps and baited them with fish heads left over from the previous day's catch. He concedes that Edwards was the superior fisherman and would usually haul in twice the number of fish for an afternoon, but when it came to shell fish the cornetist had the edge. He trapped as many as a dozen lobsters in a single night. The caretaker who picked up the garbage from their cabins noticed the lobster shells and began wondering how anyone could buy so many lobsters. The secret was out a few weeks later when he watched LaRocca running the traps at two in the morning.

In September the Dixieland Band was back in New York at still another of Busoni's ballrooms, the Danceland at 95th Street and Broadway. However, the twelve o'clock curfew effective in the city had killed off the late dance hall crowd. Musicians' salaries had been reduced commensurate with the shorter hours. Nevertheless, steady work had sharpened up the outfit and before long Seaberg and Vanicelli were helping the famous jazz band to click as it had done years before with the original combination.

On December 28, 1922, LaRocca signed a contract with booking agent J. J. Robbins, who immediately arranged for a series of recordings for the OKeh Phonograph Corporation. The Robbins contract covered recording only and in no way infringed upon the legal arrangements with Sixti Busoni.

The first four sides were cut at a single session on January 2, 1923 (see Table 7), with Don Parker added on saxophone. While the sax was forced on the Dixielanders against their better judgement for the 1920–21 Victor series, times had changed and LaRocca himself now insisted that the instrument be used. The saxophone had become the symbol of the jazz-gin-shimmy age, and no bandleader with any hope of survival dared be without one. As before, it served no good purpose in the dixieland ensemble and wrought its usual conflict with the trumpet.

"Some of These Days" is undoubtedly the liveliest of the OKeh series, with Edwards working in some odd tonal effects with his muted trombone. (Lucky possessors of this record will be surprised to learn that these effects were produced not with a conventional trombone mute, but with a special kazoo Edwards had modified for insertion into the bell of the horn!) "Toddlin' Blues," the only recorded example of this LaRocca tune, appears on the reverse side. Edward's army experience must still have lurked in his mind, for his trombone quite unexpectedly does a jazzed-up version of the mess call as its solo contribution, minus accompaniment. "Barnyard Blues" is nearly identical with the earlier recordings, except for being rendered in a much faster tempo. Seaberg, obviously coached by LaRocca, plays the Shields clarinet part faithfully. Perhaps the most enjoyable feature of this newer version is the strong ensemble piano work of Henry Vanicelli, who plays a solid four-four much in the style of his predecessor, Frank Signorelli. "Tiger Rag" moves along at its usual pace, with Don Parker volunteering saxophone breaks where once noodled the clarinet of Larry Shields.

The records were released on March 1, but the OKeh series came to an abrupt halt when LaRocca discovered that they were being advertised by the company as "race" records. Although the white Original Memphis Five had recorded under the names of "Ladd's Black Aces" and "The Cotton Pickers" in order to capture a large share of the colored market, the Original Dixieland Jazz Band would have no part in this kind of masquerade.

After a season on Broadway, the band was ready for a tour of the New England states under the auspices of the Victor Company, beginning on April 14, 1923, with a Society Tea Dance at the Copley Plaza Hotel in

Boston. That afternoon radio station **WNAC** made the first broadcast of a jazz band. The Boston *Herald* of April 8 announced the event:

New England radio fans, and especially those who love the jazz, have a big treat coming through the ether between 4 and 5 o'clock on Saturday afternoon, April 14, when the Original Dixieland Jazz Band, the favorites of two continents, will play their own incomparable melodies at a society concert and tea dance at the Copley-Plaza Hotel in Boston. One of the features that will be broadcast from Boston on that date will be the latest dance sensation of Broadway, and the first time played outside of New York, "The King Tut Strut," recently composed by D. J. LaRocca, the leader of the Original Dixieland Jazz Band. Radio operators, fans and jazzers should be at their properly attuned receivers at 4 o'clock on above date, as the first radio "Jazz Jubilee" will be well worth listening to, and if amplifiers are used it will be possible to foxtrot to the latest Broadway favorites.

The New England tour met an untimely and tragic end on April 17 with what the musicians now refer to as the "Bangor Incident." The band, advertised in the Bangor, Maine, *News* as "all white and gentlemen," had been engaged for a dance at the auditorium that Tuesday night on a $750 guarantee plus 50 per cent of the coatroom and concessionaire privileges. The dance was a huge success and was terminated by the feature of the evening, a fox-trot contest for the handsome Dixieland Trophy. LaRocca, trusting dance managers about as far as he could throw a Steinway, always had an assistant on the payroll whose sole duty was to stand at the entrance with a mechanical counter in his hand and keep track of total attendance. After the dance, when the musicians went to the manager's office to collect their fee, they were handed a flat $750 in cash with no cut on concessions. The story from here on has two versions: according to LaRocca, when he demanded his 50 per cent cut the manager brandished a revolver in his face and told him to hit the road. The other four musicians scattered like scared rabbits, he says, leaving him alone to hitchhike back to New York with $750 cash in his pocket. Edwards scoffs at the idea that he and the boys left Nick stranded. He explains that when they asked for more money the managed laid two Colt .45's on the table and remarked, "This is all we've got left." The musicians took this as a subtle hint, he says, and lost no time finding the nearest exit.

But whatever their mode of transportation, whatever their momentary disposition, the retreating troops of the Original Dixieland Jazz Band eventually found their way back to the big city and, probably after exchanging a few cutting innuendoes, settled down to the less lucrative but far more reliable pursuit of furnishing jazz to the Broadway dance palaces. Yet even this stand was to prove untenable as commercial forces in the ever-changing entertainment world closed in to deal their final, fatal blow.

For LaRocca, the end was near.

15

Fin de Siècle

When the vigilance societies and lawmakers had successfully completed their job of outlawing the shimmy and banishing jazz from all respectable places of entertainment, a new void was created in the music world. Violins and 'cellos returned to the pits of Broadway theaters, and no first-rate hotel or restaurant dared hire a jazz band. Meanwhile, the Victor Talking Machine Company set an example for the whole phonograph industry by steadfastly refusing to record any more jazz music. The unmentionable four-letter word henceforth disappeared from Victor labels.

Ironically the city that had been first to recognize jazz was first to become its graveyard. White musicians of promise quickly abandoned jazz as a washed-up fad in their frantic attempts to save their musical careers. Those who could read music—and most of them could— went along with the nationwide return to large note-reading orchestras, safe from the fatal association of "low-brow" music. Died-in-the-wool jazzmen who could not read music quickly became orphans in an unfriendly city, while the New Orleans boys who played from the heart rather than from paper went back south to their homes and subsequent obscurity.

In Harlem, where the new laws were not enforced, jazz still carried on. Strange as it may seem today, the

Original Dixieland Jazz Band had been instrumental in introducing jazz music to that section of the metropolis. The pages of Variety tell the story. Long before the appearance of Duke Ellington and Fletcher Henderson, long before King Oliver came north, the Dixieland Band was playing vaudeville bills in Harlem. These appearances continued after the band's return from England, and as late as 1922 they were still creating a sensation in the colored colonies of uptown Manhattan. The theatrical section of the New York *American* for Sunday, January 15, 1922, published photographs of celebrated artists of the day then performing at Harlem theaters. Among those shown were Vie Quinn, Eva Le Gallienne, and Nick LaRocca.

And so now, in 1923, with no jazz to be heard on the stage, in the ballroom, or on phonograph records, it appeared to the crusaders that the dreadful musical scourge they had worked so hard to stamp out had finally been wiped from the face of the earth. The death of jazz was boldly announced:

Jazz is dead! Peace to its soul, though it gave little peace to others! . . .

The decline and fall of jazz, they say, has been going on apace during the present theatrical season, as attested by the success of the non-jazz musical offering in the New York theater and the comparatively short runs of the attractions featuring jazz music. But for once it seems probable that New York did not start a vogue, this particular one being the return to sane music, and especially to sane dance music.

Musicians generally, and particularly leaders of dance orchestras, are of the opinion that the march back to normalcy as regards dance music started in Boston, and with the Leo F. Reisman dance orchestra, which has been engaged to come to New York for the first time in "Good Morning, Dearie."

Two years ago in Boston, Reisman, the leader of the orchestra, was called upon to put together a dance organization for the Brunswick Hotel. Jazz was then at its height, and, aside from clarinets and trombones, the alleged musical instruments of a dance orchestra included such melody makers as cowbells, whistles, sleighbells, cocoanut shells and even tin pans and wooden rattles.

Reisman eliminated both clarinets and trombones and he informed his trap drummer that he was to play only the drums, while to the orchestra in general he issued the instruction that it was to play only the notes indicated by the score and no interpolated effects would be permitted . . . The new tempo was somewhat more deliberate than that usually set by a dance orchestra and the rhythm was rather suggestive of a glide than a hop.

"We do not depend upon our rhythm to create interest," says its leader . . . "We have found that the elimination of the clarinet and trombone have been the greatest aids in getting away from the jazz suggestion. Both instruments have their proper field, but we have been most fortunate with an instrumentation consisting of two violins, a muted trumpet, a tenor saxophone, a string bass, piano, and drums." [1]

The Original Dixieland Jazz Band, with a steady job and a fair following, were perhaps still unmindful of the calamity steadily overtaking them. They continued to play as only they knew how, still producing good jazz but for the most part riding along on their famous name and past reputation. With the top-paying jobs gone and the trend of dance music changing, there was certainly little incentive for new jazz compositions. A 1923 creation of Eddie Edwards, "Ritarding Cheese," never played and never published, remains as a grim reminder of those declining years.

The Dixieland Band swapped places with the Mem-

[1] From an unidentified clipping in the LaRocca Collection.

phis Five during the winter of 1923–24, LaRocca's band returning to the Balconades and the Signorelli combination opening at Busoni's dance hall in Brooklyn for the season.

Then, on February 12, 1924, there occurred the most momentous musical event of the decade. It was on that afternoon of Lincoln's birthday that Paul Whiteman, weighing over three hundred pounds and dressed in faultless evening attire, mounted the podium at Aeolian Hall in New York City and raised his baton on a new era in popular music. It was the much publicized "Jazz Concert," paradoxically enough, that sounded the death knell for jazz.

Paul Whiteman had come a long way over a strange route. Born in Denver on March 28, 1891, he had been raised in the environment of classical music. His father, who was supervisor of music for the Denver public schools, made certain that his son followed academic tradition in this field. While still in his teens Paul played first viola in the Denver Symphony Orchestra, in the San Francisco People's Symphony, and finally in the Minetti String Quartet. During the World War he directed a navy band of fifty-seven men on Bear Island, California. Subsequently he formed his own orchestra at the Potter Hotel in Santa Barbara, and, after hiring pianist Ferde Grofé as arranger, introduced "symphonic" jazz at the Los Angeles Alexandria Hotel in November of 1919. In 1920 he brought his nine-piece combination east for a job at New York's Palais Royal Café. His new ideas for "dressing up" jazz were everywhere warmly received, and after returning from a tour of England in 1923 he began earnestly rehearsing his orchestra for the realization of his lifelong dream: a jazz concert at Aeolian Hall, that rock-bound stronghold of classical music.

The concert, advertised as "An Experiment in Modern Music," was a most daring and audacious undertaking, representing an investment of eleven thousand dollars by the enterprising Mr. Whiteman. There had been strong doubts from the very beginning that the staid, regular patrons of Aeolian Hall would take the trouble to attend any such unabashed demonstration of music that had already been condemned as sheer discord, by whatever name it might now be called. One can almost feel the anxiety of Paul Whiteman on that memorable afternoon as he stepped up to the podium and looked out for the first time upon an audience composed of such musical dignitaries as Walter Damrosch, Leopold Godowsky, Victor Herbert, Jascha Heifetz, Fritz Kreisler, Sergei Rachmaninoff, Leopold Stokowski, and Josef Stransky. In his own book, *Jazz* (New York, 1926), written while the Aeolian experience was still fresh in his mind, Whiteman describes his anguish: "I believed that most of them had grown so accustomed to condemning the 'Livery Stable Blues' type of thing, that they went on flaying modern jazz without realizing that it was different from the crude early attempts."

To show how far American music had "progressed" in seven short years, he began the program with La-Rocca's "Livery Stable Blues," played exactly as the Original Dixieland Jazz Band had played it in 1917. The soaring clarinet of Gussie Mueller was heard for the last time as five men from the Whiteman orchestra paid one last fleeting tribute to the memory of pure jazz. The ovation that followed was a frightening surprise to

the bulky bandleader. "The audience listened attentively
to everything and applauded whole-heartedly from the
first moment," he continues in his book. "When they
laughed and seemed pleased with 'Livery Stable Blues,'
the crude jazz of the past, I had for a moment the pan-
icky feeling that they hadn't realized the attempt at
burlesque—that they were ignorantly applauding the
thing on its merits."

Then, as the five men disappeared into the sea of
tuxedos that engulfed the stage, the entire Whiteman
ensemble gave forth with the new sound in American
music—"Symphonic Jazz." Simple hack tunes such as
"Yes, We Have No Bananas" and "Alexander's Ragtime Band" were embellished with impressive crescen-
dos and sudden changes of tempo, as hashy brass
sections alternated with sluggish violins amid the mounting rumble of kettle drums. Flutes, oboes, bass clarinets,
four sizes of saxophone, octavions, celestas, flugelhorns,
euphoniums, basset horns, French horns, and a bass
tuba augmented the traditional jazz instrumentation to
usher in the era of "sweet" dance music that was to
dominate the entertainment world for nearly two decades. Among the twenty-three men who "doubled" on
thirty-six instruments were such famous names as pianist
Zez Confrey, composer of "Kitten on the Keys," and the
versatile ex-bandleader Ross Gorman, who kept busy
with ten different reed instruments.

It is no wonder that New Orleans clarinetist Gus
Mueller became seasick and had to head for the wings.
Whiteman, in his book, relates his experiences with
Gussie:

Gus Mueller . . . was wonderful on the clarinet and saxophone, but he couldn't read a line of music. I tried to teach him, but he wouldn't try to learn, so I had to play everything over for him and let him get it by ear. I couldn't understand why he was so lazy or stubborn or both. He said he was neither.

"It's like this," he confided one day. "I knew a boy once down in N'Awleens that was a hot player, but he learned to read music and then he couldn't play jazz any more. I don't want to be like that."

A little later, Gus came to say he was quitting. I was sorry and asked him what was the matter. He stalled around awhile and then burst out:

"Nuh, suh, I jes' can't play that 'pretty music' that you all play. And you fellers can't never play blues worth a damn!"

The extreme in symphonic jazz was reached later in the concert when guest artist George Gershwin sat down at the piano for the first public rendition of his new concerto, *Rhapsody in Blue,* accompanied by the orchestra in an arrangement by Ferde Grofé. To many, the theme sounded suspiciously similar to "Limehouse Blues," played earlier on the same program.

The Whiteman concert marked the end of small jazz combinations, as music became sweeter and dance managers, booking agents, and theatrical producers demanded large "symphonic" type orchestras. The orchestras of Paul Whiteman, Nat Shilkret, Horace Heidt, and George Olson set the pattern on Victor records, while Leo Reisman and Harry Reeser followed suit on radio.

But the Dixieland Band played on, still pushing out jazz in its pure form, still fighting the trend. The Charleston craze had started, and for awhile things looked better. Dancers who never could shimmy had no trouble mastering the new step—providing they could stand the pace. The furious, continuous round of parties—hooch, women, and song—continued for the Dixieland Band and their loyal following. LaRocca wore two-hundred-dollar suits fashioned by Manhattan's smartest tailor and diamond rings that sparkled in the glaring lights of nocturnal Broadway. In the afternoon he could be seen heading for Long Island in his fire-engine-red Stutz Bearcat. On the Long Island Speedway you could pay a dollar for the privilege of driving as fast as you liked, and LaRocca, garbed in goggles and duster, with his cap fitted backwards, tried to make his Bearcat fly. Whatever his critics may have said of him, Nick LaRocca was part of the American scene, a participant in the Jazz Age he had helped name.

Yet the fast life was not a happy one for the Dixieland bandleader. Music and women threatened to drive him to madness, while friction within the band made every performance a trial. Edwards was clowning, he claims, and every direction or suggestion he made was met with a juicy "razzberry" from the mischievous trombonist. Life had become an unbearable nightmare for LaRocca.

Then, one night in January, 1925 came the collapse.[2] Nick LaRocca, suffering a complete nervous breakdown, could carry on no longer and was advised by his physician to give up music forever.

Leaving the Stutz Bearcat in a garage on 145th

[2] The approximate date is established through the correspondence of LaRocca (in New Orleans) and Edwards (in New York) between the dates of February 27 and March 26, 1925, relative to the collection of back salary from Sixti Busoni, owner and manager of the Balconades. Edwards and LaRocca refused to accept a settlement of $1,000 secured by Local 802, and Busoni died in May, 1927, still owing the Original Dixieland Jazz Band $2,015.68.

Street, he piled a few belongings into his other car, a 1923 Buick sedan, and headed south. Half-crazed by the music that still rang in his ears, he followed no signs, consulted no maps. Time was of no significance to the sick bandleader, for it had suddenly ceased to move. Heading south and going home was all that mattered. He arrived home after an incredibly long period and via a fantastically indirect route.

Jazz was dead, from the standpoint of a purist, and with it was gone the spirit of the Original Dixieland Jazz Band.

16

The Lean Years

Crippled by the loss of its great cornetist and leader, the Original Dixieland Jazz Band quickly vanished into obscurity. Eddie Edwards left almost immediately, forming his own band for a job at Busoni's Rosemont Ballroom in Brooklyn. Only one original member was left now—Tony Sbarbaro—and he was determined that the Dixieland carry on. Under his leadership the band opened in New York at the Cinderella Ballroom, Broadway and 48th Street, with a new line-up. Tall, youthful Artie Seaberg was still on clarinet, but Wilder Chase had replaced Henry Vanicelli at the piano, Harry Gluck took over the trumpet spot formerly occupied by Nick La-Rocca, and a trombonist by the name of Kaplan came all the way from China to replace Eddie Edwards. It was Wilder Chase, a close friend of LaRocca, who had been instrumental in getting Seaberg into the band back in 1922.

An embarrassing situation existed for a few weeks when the bands of Eddie Edwards and Tony Sbarbaro were both billed as "The Original Dixieland Jazz Band." The conflict was settled by a letter from LaRocca to Sbarbaro, dated March 30, 1925, in which the ex-bandleader assigned full rights to his former drummer for use of the band's name:

. . . If you think it advisable to hire a lawyer and you
would want to pay the expenses, I will give you all the support
you want in this matter, as I am sole owner of the name as it
stands and gave only to you the privilege to use same under certain conditions. Of course, for me to stop him [Edwards] and
spend my money would not mean anything for me . . . If you
care to seek legal protection, I am here to back you up, and
Toney I do not ask one penny from you and am only glad to
see you get along. When you do see the lawyer, explain the
matter that I am the owner of this name and give only you the
permission to use same . . ."

Eddie Edwards changed the name of his band to
"The Southerners" and for a while Tony's problems
were over. However, it appears that Harry Gluck had a
previous engagement and was able to remain with the
Dixieland Band only a few months while Tony searched
desperately for a qualified replacement. It was about
this time that Artie Seaberg heard rumors of a high
school boy who had been starting fires with his cornet
over in Brooklyn, so Artie made haste to investigate. The
inspired youth was Henry "Hot Lips" Levine, a name to
achieve coast-to-coast fame over the radio networks in
much later years but, because of an unusual set of circumstances, probably one of the most underrated jazzmen in history. It is not so much Henry Levine's work
with the Original Dixieland Jazz Band, which was certainly far past its zenith when he joined, as it was his
outstanding accomplishment of keeping dixieland jazz
alive during its darkest years that entitles him to a prominent place in any book purporting to a complete account of jazz history. Nevertheless, there is no doubt that
Levine's memorable experience of playing with such
greats as Seaberg and Sbarbaro during the most impres-

sionable years of his life was of prime significance in the moulding of his musical personality.

Born in London, England, on November 26, 1912, Henry arrived in the United States at the age of six months. Like the jazzmen of old, his talent for music was manifest at an early age. Before he had reached his eighth birthday he had started cornet lessons and was receiving a thorough groundwork in theory and harmony under the able supervision of Max Schlossberg. But it was doubtless a certain evening in 1922 that left its indelible mark on his young musical mind. The precocious eight-year-old had been taken by his father to a café called Stauch's on Coney Island, where he listened, wide-eyed, to the most thrilling music he had ever heard. The great moment is best described in Henry's own words:

... Two bands were playing there, one large brass band on the balcony called Eppy's and a five piece jazz band called the Original Dixieland Jazz Band on the ground floor. When I heard this outfit, I was enthralled and decided that this was the music for me. I had been playing cornet and studying legitimate music, but until that night I had not been excited about trying to play dance music. After that evening in Stauch's, I had no more doubts, it was going to be dixieland or nothing. Nick LaRocca was playing the cornet left-handed, holding the horn in his right hand and fingering the valves with his left. Shields seemed to hold the entire clarinet mouthpiece in his mouth and got a peculiar kind of flat tone. But the band had the sound, a happy sound with the fine two-beat, and this was it . . .[1]

Henry never forgot his first contact with pure jazz. The "happy sound" echoed in his young head for years afterward and gave him no peace until he could duplicate it on his own horn. School work occupied most of

[1] From a letter to the author dated March 21, 1958.

his time—Henry was inclined to be studious, but evenings gave him the chance for a limited amount of club
work with small pickup bands that tried, without much
success, to play jazz.

Then, in the spring of 1926, came the second great
event that helped shape his career. While playing a
Friday-night job he was introduced to Artie Seaberg,
who took an immediate interest in the fourteen-year-old
cornetist. He seemed especially to like the tone that
Henry evoked from his insrument—a tone that was
something like LaRocca's—and thought he could learn
to drive a jazz band. Seaberg thereafter called frequently at the home of the high school sophomore, bringing along his clarinet and a pile of phonograph records
of the Dixieland Band. With great patience he proceeded
to teach Henry "Tiger Rag," playing each part over and
over on his clarinet to get across the idea of "drive"—
when to hit hard on the downbeat, when to drop out to
let the clarinet and trombone come through, how to lead
into a chorus—all the tricks that Artie could explain
only by doing. Although jazz cannot be taught and must
be assimilated, the young cornetist caught on quickly
and was more than ready when his big opportunity came
—the chance to try out for the Original Dixieland Jazz
Band. When Harry Gluck left, Seaberg introduced the
lad to Sbarbaro, who hired him after one trial chorus.
Henry continues:

. . . Since I could read music quickly, Artie thought that we
could play brand-new tunes and I could blow the lead the first
time through. This was for me the crowning achievement of
my young life and it worked out just as Seaberg thought. On
the current tunes of the day, I would play the melody first time

through with Sbarbaro beating out the time. By the time the third chorus rolled around, the whole band would play these new tunes as if they had been playing them for years. These were musicians with such good ears that they learned new tunes almost at once. At the same time, I learned their songs, such as "Barnyard Blues," "Fidgety Feet," "Bluin' the Blues," and dozens of others . . .[2]

The Cinderella engagement lasted about seven weeks, after which the band moved to the Paradise Ballroom in Newark for a somewhat shorter stay. But times were tough for small combinations and the Dixieland Band soon found itself hopelessly fighting a lost cause. When it collapsed early in 1927, Henry returned to New York to join the orchestra of Vincent Lopez at the Casa Lopez on West 54th Street. Many years of big band work followed, during which he played with the orchestras of George Olson and Rudy Vallee, and in the pits at such historic Broadway musicals as *George White's Scandals, Roberta, Life Begins at 8:40,* and *Hold Your Horses*— besides finding time to attend the City College of New York.

Meanwhile, Tony Sbarbaro reorganized the Original Dixieland Jazz Band at the Rose Danceland on 125th Street, increasing its size to ten men and employing written arrangements. But competition was keen among the bigger outfits, so at the end of the winter season Tony, who had stuck it out to the bitter end, finally decided to disband the ponderous aggregation which was by now dixieland in name only, and earn his living jobbing around with other bands at various Manhattan night clubs.

[2] *Ibid.*

Yet, despite public disinterest in hot jazz, a few small outfits continued to gather for recording sessions. These were the days when pickup bands were literally pickup bands. The advent of sound movies had thrown hundreds of pit musicians out of work all over the city, and an enterprising jazzman could form a small combination by standing at the corner of Broadway and 47th Street and taking his pick of the tuxedoed union men who wandered by, instrument cases in hand, seeking jobs and ready to play at a moment's notice.

One of these transient jazzmen was a popular cornetist named Leon Bismarck Beiderbecke, known simply as "Bix" to his many friends in the jazz underground. Bix and Artie Seaberg palled around together, being about the same age, and many were the times Artie paid Bix's train fare back home to Davenport, Iowa, when the cornetist was flat broke from lack of work.

Like most of the promising young musicians of his generation, Bix had first heard jazz on the famous Victor records of the Original Dixieland Jazz Band. His mother recalls how he would wait until the family had gone out for an evening, then sneak downstairs in his pajamas to play his cornet to the accompaniment of the Dixieland's records, usually "Tiger Rag." Young Bix's reverence for LaRocca began at an early age and was a definite influence on his great cornet style. He had first come to New York in 1918, when he was a mere lad of fifteen, for the sole purpose of hearing his idols in person and watching their leader. LaRocca remembers being called to the front door of Reisenweber's one night to authorize the admittance of the youngster, who had tried to gain entry by posing as the bandleader's son! Bix hung around for a week, staying at LaRocca's hotel suite during the day and sitting quietly at a café table at night, hypnotized by the music he loved. Then LaRocca became worried over the possible consequences of harboring a minor and shipped the boy back to Davenport. This vivid boyhood experience must have carried its effect throughout all of the musician's life.

It was not until five years later, while the Original Dixieland Jazz Band was playing at the Balconades, that this same young man, now fully mature and rather pleasantly inebriated, walked up to the bandstand to renew his acquaintance with LaRocca. He and George Brunies had been playing around Chicago with the Wolverines, occasionally dropping in at the Gennett studios in Richmond, Indiana, to record such LaRocca tunes as "Sensation Rag" and "Lazy Daddy."

Now, in 1927, the combination known as "Bix Beiderbecke and his Gang" assembled at the OKeh recording studios in New York to produce what most experts consider the most outstanding jazz of the late twenties. With Frank Signorelli (the last great pianist of the Original Dixieland Jazz Band) on piano, the Beiderbecke Gang inaugurated the historic series with a rousing number that seemed to recall the ghosts of its retired composers, Nick LaRocca and Larry Shields: "At the Jazz Band Ball." Even Adrian Rollini's rollicking bass sax went in for a chorus. But Frank and Bix, who liked to eat as well as anyone, could no longer resist the hundred-dollar bills being waved in their faces by Paul Whiteman; they subsequently joined the vast Whiteman empire. It was in 1928, at Carnegie Hall, that the Whiteman orchestra introduced Frank's first song,

"Gypsy," and a few years later, at the Metropolitan
Opera House, his well-known "Park Avenue Fantasy."
Other Signorelli hits followed—"A Blues Serenade,"
"I'll Never Be the Same," "Stairway to the Stars,"
"Anything," and finally "Sioux City Sue." A wisecracking kid named Bing Crosby was doing his "vo-dee-
oh-do" with Whiteman's vocal trio—The Rhythm Boys
—in those days; scattered throughout the twenty-
seven-man "symphonic jazz" orchestra were such yet-
unfamiliar faces as Tommy Dorsey, Jimmy Dorsey, and
Frankie Trumbauer. In the meantime, farther west, a
grinning clarinetist known as Benny Goodman had graduated from the Ted Lewis band and was joining trombonist Glenn Miller in the Ben Pollack group. These
were all real jazzmen suffering from the blight in their
trade, genuine creative musicians who would rather
"swing out" than glue their eyes to a staff, and who were
patiently awaiting the new jazz age that was sure to
come.

But although such talented jazz figures as Bix, Signorelli, and the Dorsey brothers were temporarily neu-
tralized by Paul Whiteman, a determined Red Nichols
insisted that these men be heard. Between 1928 and
1930 his Five Pennies boasted—at one time or another
—Jack Teagarden, Tommy Dorsey, Glenn Miller, Miff
Mole, Benny Goodman, Pee-Wee Russel, Wingy Ma-
none, and Gene Krupa. However, even Red was to lose
the battle. By 1931 he was recording for Victor with
a ten-piece orchestra which found Tommy Dorsey and
Glenn Miller side by side in the trombone section, with
Benny Goodman blowing clarinet and Gene Krupa beating the drums! It was these men, all within a few years

of one another in age, who were to spearhead a new movement in hot music in another decade.

If anyone stopped to remember, the Original Dixieland Jazz Band seemed far in the past. Johnny Mercer wrote a popular ditty called "The Story of the Dixieland Band" during the early thirties, its last stanza proclaiming that the musicians were now up in heaven "and a-playin' away," so it was clear that the famous band of that name had by now passed into the realm of popular mythology. But its members were far from heaven and, because music could no longer keep them alive, were not "a-playin' away." Tony Sbarbaro claims to be the only member who kept musically active during all of the treacherous early years of the depression, and there is evidence that even he was compelled to fill in with other lines of endeavor.

Larry Shields' movie career was short-lived, but he managed to keep going for a few more years around Hollywood at such places as the Tent Café, Sunset Café, 400 Club, and the Vernon Country Club. As times worsened, Larry's lapses of unemployment become longer and more frequent. In desperation he returned to the city of his birth in hopes that friends could help.

In New York Eddie Edwards survived for awhile with his Silver Slipper Orchestra. In the year of the historic stock market crash he entertained a lively crowd at Roseland and played for a dance marathon at Madison Square Garden, broadcast locally over WMCA and WJZ. But the job at Krueger's Auditorium in Newark on New Year's Eve of 1929 was one of the last. The great tailgate artist of the Dixieland Band then gave up music for a more steady line of business, opening a magazine stand on Seventh Avenue across from Carnegie Hall. Eddie claims that he had a nice little business going there and takes issue with Walter Winchell and other gossip columnists who painted pathetic pictures of the forgotten trombonist "selling newspapers on Columbus Circle."

Then, in the fall of 1933, Eddie had an unusual caller. Brad Gowans, a hustling Yankee with a dixieland complex, had come down from Boston full of youthful enthusiasm and new ideas for starting a revival of dixieland jazz. Although Brad is today noted as a jazz trombonist, he was long an admirer of Larry Shields and schooled himself along these lines on the clarinet, developing a tone very much like that of his master. He had in mind a reorganization of the Original Dixieland Jazz Band. Although Edwards and Sbarbaro were dubious, they were willing to experiment. Repeal of the Eighteenth Amendment was about to take effect, ending fifteen dry years of Prohibition, and Brad wanted a five-piece band to play on the Park and Tilford boatload of liquor coming into New York Harbor on December 6. The publicity value of the stunt could hardly be overestimated, especially considering nationwide newspaper coverage, Movietone newsreel shots, and a coast-to-coast radio hook-up. But the plan suffered a sudden death when the two largest networks declined to broadcast liquor commercials.

Brad then tried in vain to talk Eddie and Tony into a series of one-nighters throughout the New England states. But the two veterans of the Dixieland Band had done their share of travelling and found little appetite for the rigors of roadwork. A hotel job or radio show were more to Edwards' liking, but he feared that most people would no longer listen to five men playing i ι one style for a whole evening. With this in mind he wrote to Nick LaRocca on December 2, proposing a ten-man band in which LaRocca was to play second trumpet—a sort of "band-within-a-band" type of thing in which LaRocca would be used as the lead on jazz numbers. Meanwhile, Brad had lined up a job at the old Hammerstein Theater at 53rd and Broadway, which was then being converted into a beer garden.

But no music stirred within the heart of Dominic James LaRocca. His adjustment to private life had been effected, he found family life reassuring, the contracting business earned him a good living and he intended to stay with it.

So it was that Brad Gowans' efforts to revive jazz failed. On theater marquees from Maine to California there blazed the title of the year's biggest and most spectacular motion picture musical:

THE KING OF JAZZ
starring Paul Whiteman

17

The Comeback

Symphonic jazz—the artless product of million-dollar advertising and the American love for something big—ran its course for several years, eventually losing some of its symphonic characteristics. Orchestras became smaller and arrangements less ambitious as the Great Depression cast its shadow across the continent, but written jazz was apparently here to stay. Although Red Nichols, Frankie Trumbauer, Bix Beiderbecke and a number of dedicated small combo men fought valiantly to keep the spark of pure jazz alive during those bleak years, they were relatively unknown to the general public, and the sweet music of Guy Lombardo, Vincent Lopez, Glen Gray, and Fred Waring dominated the scene. Commercial jazz had finally become, to borrow the well-chosen words of writer J. S. Moynahan, ". . . a diluted, degenerate echo of the music that swept a country, a world."

But now the pendulum was swinging back, the cycle nearly complete. The opportunity was there, the public was ready, and Benny Goodman was the man. Although freshly graduated from the recording sessions of Red Nichols, Bix Beiderbecke, and Ben Pollack, his business sense told him he couldn't sell a small band to a big public in this age. In 1934, with these facts carefully weighed in his mind, he realized every young musician's ambition by forming his own orchestra, a modest thirteen-man group considered neither large nor small at that time. Organized expressly as a pit band for the Broadway musical comedy *Free For All*, the new outfit gained little public attention at the beginning. However, a thirty-six-week contract for a weekly Saturday evening network broadcast kept the band eating regularly, while a series of exhausting one-night stands brought the ambitious young musicians to every level between hope and utter despair. At New York's Roosevelt Hotel, for example, Guy Lombardo was hastened back to replace the failing Goodman band after the management had complained bitterly of the "loud" brasses. To certain old-timers in the music game, it may have recalled the 1917 trials of the Original Dixieland Jazz Band at Reisen-weber's in that same city.

Their fortunes were not a bit improved at a place in Denver, where a horrified proprietor fired them inside two days. Glum and nearly heartboken, the curly-headed bandleader concluded the tour at the Palomar Ballroom in Los Angeles, half-expecting the worst. To the amazement of everyone in the band, to say nothing of critics at large, they were an overnight sensation with the Palomar dancers who, wildly cheering every "hot" number, clamored for the special Goodman arrangements they had been hearing on the Saturday night radio program. Apparently the difference in time zones had brought the music to the right people at the right time.

Thus radio had made Goodman a nationwide success, just as the phonograph had popularized the Original Dixieland Band two decades earlier. But the phonograph, which had gradually reached the status of a museum piece in the wake of radio's meteoric rise, began to stage a spectacular comeback about this time. The invention of the electrical pickup, making possible the electronic amplification of recorded music, led naturally to the radio-phonograph combination and with it a renewed interest in phonograph records. Advertising again created a new desire in the public to hear its favorite radio stars on wax, where they could be recalled anytime at will, and the Goodman orchestra grooved a series of Victor platters that sold phenomenally among the high school set. No "gate" who was worth his weight would even dream of going on a week-end trip with the crowd without packing a portable phonograph and a stack of Goodman records into the rumble seat of the old beat-up Model A.

The Swing Era had begun. Compared with the Goodman band's hectic westward excursion, the return trip was a triumphal procession. After six spectacular weeks at the Hotel Congress in Benny's native Chicago, the band continued east to New York. By the time he reached the big city, Benny Goodman was everywhere recognized as the King of Swing.

As with any great revolution in dance music, the "swing" movement was supported by the dancers it had fostered—in this case, the "jitterbugs"—those energetic teenagers who expressed their love for rhythm by throwing one another into the air. Special dance steps were invented for the new musical style, and forgetful parents who were "ballin' the jack" in their youth wondered what music was coming to when they saw their own children "truckin' on down."

It was the era of the Name Band. The original members of the Goodman aggregation deserted one by one to form their own bands—Harry James, Bunny Berigan, Ziggy Elman, Teddy Wilson, and finally Gene Krupa. Name bands reproduced like rabbits, scattering their breed throughout ballrooms, radio dance parties, and a thousand Hollywood movie shorts. On and on they marched across the American scene—Charlie Barnet, Tony Pastor, Charlie Spivak, Claude Thornhill, ad infinitum—and despite obvious attempts at stylization, they all sounded basically alike: hashy brass sections rising and sitting in unison like so many mechanical men, alternating with rows of cheek-puffing sax men who did the same. The credit belonged to the arranger and the publicity agent.

But this mad rush to conformity was not without its blessings. With the advent of swing, and thanks to the daring of Benny Goodman, solo improvisation was again publicly acceptable. Jam sessions came out from behind locked doors, and the Goodman type of trio or quartet— the single wind instrument (preferably reed) leading a battery of rhythm instruments covering the alphabet from aeolians to xylophones—again offered opportunities for creative jazzmen to make themselves heard.

In the Goodman group it was Benny himself who awed his fervid clan with flexibility and classic technique previously unheard of in such large doses. The swing style of drumming was also spectacular, if totally unartistic. Its pioneer, Gene Krupa, who highlighted the trio specialties with furious sixteen-bar drum breaks, wandered so far from the basic beat that he sounded

more like a well-equipped kettleware salesman falling down four flights of stairs. From then on the
perspiration-drenched drummer with the face contorted
with a strange mixture of pain and ecstasy became the
trademark of high school and college swing bands from
coast to coast. History repeated itself as noise became
the means and end in popular music.

But if the swing drummer was a brand-new phenomenon, the swing clarinetist was not. Listening to
Benny Goodman, it was only a matter of time before
somebody remembered the great Larry Shields, father of
the noodling clarinet. Those "licks" of Goodman's, while
breathtaking and original in their context, echoed in
style the fifty-one breaks of Shields in the Original
Dixieland Jazz Band's historic "Ostrich Walk"—and the
modest Goodman was one of the first to admit it. In the
Christian Science Monitor, July 11, 1938, Benny wrote:
"I was playing jazz on my clarinet when I was eight
years old, listening to the records of the Original Dixieland Jazz Band, which made a terrific impression on
me . . ."

Then, in Hollywood, somebody decided to do something about it. An old friend and loyal fan of the Dixieland Band, Lew Gensler, was lining up talent for a movie
spectacular to be called *The Big Broadcast of 1937.* In
1917 Gensler was a Packard salesman on Columbus
Circle, just across from Reisenweber's, where he was a
regular patron. Also a talented pianist, he had once tried
without much success to teach the Dixieland Band one of
his own compositions. Now, nearly two decades later,
the popularity of swing music compelled him to investi-

gate the whereabouts of the band that had been the forerunner of swing. He immediately asked New York agent
William Morris to locate his friend, J. Russel Robinson,
then engaged in radio work for NBC. Robinson, excited
over the idea of reorganizing the Original Dixieland
Jazz Band for a top Hollywood musical, quickly secured
the address of D. J. LaRocca.

At the time, LaRocca was a one-man contracting
company in New Orleans, building houses and doing all
the carpentry work himself. When William Morris approached him with a contract for *The Big Broadcast of
1937*, he had two houses under construction and was unable to leave the city. Consequently, he turned down the
offer.

Nevertheless, the incident proved to Nick LaRocca
that in ten years the world had not entirely forgotten the
old Dixieland Jazz Band. The more he turned it over in
his mind, the more he sensed a growing awareness of his
unique place in jazz history. The old bitterness for music
had finally worn off, his Broadway wounds had healed,
and although he had not dared listen to music in more
than a decade, he now tuned in the radio to find out just
what was going on in the outside world.

As his radio searched the spectrum, he heard for
himself the same "riffs" and "licks" that had been
played by the Dixieland Band in those threatening days
immediately preceding World War I—those familiar
musical phrases that were considered "corny" during
the Whiteman era—but which were now being proudly
executed in all seriousness by swing band soloists over
the airwaves. He snapped off the radio with a determined flip of the wrist. His mind was made up.

His first step was to find Larry Shields, for LaRocca was a shrewd enough observer of the popular scene to realize that the whole swing movement revolved around the King of Swing himself, Benny Goodman. Therefore, if he wanted to prove to a new generation that this swing stuff was really old jazz in new dressing he would first of all need a clarinetist who could compete with the King on his own ground—and only one man on earth was capable of this feat—Larry Shields.

Although more than fifteen years had elapsed, finding Larry was no great problem. It was only necessary to go into the Shields neighborhood, inquire among the vast Shields clan, and somebody was sure to know the whereabouts of Uncle Larry.

Nick found him working in a Bible house, right there in New Orleans. Although he was a mere forty-three, his hair had turned a snowy white. As for music, he hadn't touched a clarinet in ten years. "I don't know if I can still cut it," said Larry, "but I'll try."

They practiced together for weeks. Then, on May 19, 1936, the two of them made an unexpected appearance at the Absinthe House in New Orleans. Mel Washburn, columnist for the New Orleans *Item,* was in the audience that night and described his thrill in the issue of May 23:

Tuesday night I was sitting in the Absinthe House watching the floor show . . . "And now," said Eddie Barber, the master of ceremonies, "we've got a real treat for you . . . two boys from the most famous band in the world, Larry Shields and Nick LaRocca of the old Dixieland Jazz Band." What's this, says I to myself . . . what kind of sandy are they putting over . . . and then onto the floor comes Nick and Larry . . . in

the flesh . . . and they swung into one of their old favorites.
Boys and girls THAT was swing music . . . just a trumpet and
clarinet . . . but how those boys went to town. Maybe they have
had a twelve-year layoff . . . but you'd never believe it listening
to 'em . . . and you should have heard the storm of applause
they provoked from the pop-eyed audience . . . pop-eyed because they couldn't believe that a white-haired grandpop like
Larry could make that clarinet screech, wail and sob as it was
doing. . . . or how that left-handed trumpeter could get such a
seductive swing into the rhythm as Nick was doing. Later I
asked 'em how come . . .

"We're not working here," explained Nick, "but we come
down here two or three times a week and go into the show, just
to get in tune again . . . get our lip up and wear off the rough
edges that the last 12 years have given us."

Obviously, they could still "cut" it. The word
spread quickly, and before long the retired bandmaster
was receiving offers from London, Paris, New York, and
Hollywood. The next question was, could the other
former members be persuaded to reorganize? Without
the originals, it just wouldn't be the same old Original
Dixieland Jazz Band.

Weeks of correspondence followed, letters to Edwards and Robinson in New York, laboriously pecked
out with one finger by Nick on his old typewriter. Both
were interested. Edwards had given up music when the
bottom fell out of jazz years before and was now coach-
ing a boy's baseball team at the YMCA on West 63rd
Street, working a few hours each afternoon. On June 8
he wrote as follows:

I haven't touched the trombone for a little more than a
year, except around last New Year's Eve, when I had an engagement at a country club. I tried to get into shape in seven
days, but after the second I gave that up, knowing what a New

Year's Eve society engagement was for work. I therefore
picked up the fiddle, directed and looked wise and made three
hours overtime which I could not have done on the horn, as my
lips would have been swollen ten times their size after one
hour's work.

Tony has been after me to practice up, and while it is
difficult to do that I have given some time and thought to it
. . . I worked out on a job with Tony last Friday night and
was really surprised at myself because tones seemed to flow
pretty easily and my lip did not feel much swollen after it. I
feel confidently that I could get back in 10 days as good as I
ever was, but that would have to be doing some dance work,
not sitting in a room blowing exercises.

. . . Glad to hear Shields will join and that he is back to
his old form. His stuff is still good and away above all the junk
musicians squeak through those Long Island duck dinners (on
account of the quack, quack they blow).

. . . Well, if anything comes of the movement, I will be
ready and will take a chance. That's OK by me.

Tony Sbarbaro was the only member of the Original Dixieland combination who had remained active in
music during the lean days of the depression. However,
Edwards had expressed some concern over the fact that
Tony had recently taken a band to Princeton for a week
end, representing it as the Original Dixieland Jazz Band
and allowing a second-rate trombonist to autograph
dance programs as "Eddie Edwards." He further
warned that the real Original Dixieland Jazz Band
would have trouble booking good jobs in the New York
area if Tony were not restrained from taking what he
called "doggie" jobs.

While LaRocca and Shields practiced together in
New Orleans, Edwards began a steady program of building up his lip for the day when the five members would

combine for final rehearsals in New York. But the other tenants in Edwards' apartment building were night shift workers who slept until noon, allowing Eddie only about an hour of practice daily.

As two more months passed, J. Russel Robinson, who had been surveying the band's prospects in New York and making a few preliminary contacts, urged immediate action. On June 26 he wrote to LaRocca:

Dear Nick,

Your letter of the 22nd reached me this morning, and many thanks for same . . .

I hate to see the whole month of July go by without your being here, because there are many things of a business nature regarding the band's future that should be looked into and decided upon. Then too, we should have the affairs of the band in some management's hands for at least six weeks before we intend working, so as to give said management time to procure the proper kind of engagements for us.

As I said before, the Music Corporation of America is interested, and they are the biggest and best bet of all managements. But Billy Goodhart, one of the heads, said he would want to hear the five of us before talking business. I really think we'd have to have a lot of lucky breaks in order to get working by September 1st, if you wait until August 15th before coming. But of course if it is impossible for you to come sooner, it will have to be that way. I know Tony feels the same way I do about delay. You understand, of course, that all bookers work very far in advance, some dates being made even six months in advance, with the modern way orchestra booking is handled.

I saw Abe Olman, head of Feist's, yesterday and he told me he is anxious to get together with you regarding the reissuing and bringing out modern arrangements of all the old Dixieland numbers. He seemed to think it a marvelous idea that we were going back into the game and predicted big things . . .

Sincerely,

Russel

Shields and LaRocca, always the perfectionists, now stepped up their practice schedule in earnest. Soon they were playing together nightly, dropping in on night clubs, parties, and miscellaneous social events, trying to see how many jobs they could fill in the course of a week. But two incidents nearly spelled doom for the comeback attempt. LaRocca, exhausted from working days as a carpenter and playing music all night, fell asleep at the wheel of his automobile on the way home one night and crashed into a tree. Fortunately, a single day's hospitalization was all that was required. Then, according to La-Rocca, the New Orleans musicians' union charged him with sitting in with nonunion bands and brought pressure on New York's Local 802 in an effort to keep him from playing professionally in that city.

Neither of these obstacles was sufficient to break up the reorganization, however, and early in July Nick and Larry shoved off for New York. They quickly located the remaining members and prevailed upon them to quit their present jobs. Although Robinson and Sbarbaro subscribed to this idea immediately, Edwards, still doubtful about the band's future, remained uncommitted. Also, as in the past, he was reluctant to recognize LaRocca as leader and exclusive manager for the organization. On the written agreement drawn up by the members, his signature is conspicuously absent.

J. Russel Robinson, through his connections at NBC, arranged for afternoon use of a rehearsal studio. In giving an account of their first reunion, LaRocca relates that everyone was present except Edwards, who had decided at the last minute that he was unable to take time off from work. So after a brief wait the bandleader jumped into a taxi, raced over to the YMCA, and did some persuasive talking to bring the balky trombonist back.

The first sounds of the new Dixieland Band were hardly encouraging. Edwards' lip gave out after the first twenty minutes, the penalty for a year's inactivity. Shields seemed despondent and generally in a trance, possibly discouraged by the failure of the five pieces to work together. Robinson was tolerant but detached, uninspired. They tried mutes, argued over the advisability of adding a bass player. The first session ended with a few threats and many doubts.

By the end of the third rehearsal, however, a big improvement was evident. When the three wind instruments achieved balance, the band began to click. Edwards' powerful counterpoint backed up LaRocca's melody, the old wheel started rolling, and Shields, feeling the push, was soon giving out with smooth, liquid runs that flowed effortlessly through the ensemble. Three weeks of steady practice brought the band to its old peak and ready for its first public appearance in more than a decade.

The William Morris agency had arranged a debut on Ed Wynn's weekly radio broadcast over the NBC Red Network on the evening of July 28. A few days before the program, Robinson accompanied LaRocca upstate to Wynn's home, where the famous comedian perused Nick's scrapbook with great relish as he collected ideas for the script.

On the following Tuesday night, the Original Dixieland Jazz Band faced the largest audience in its history, a hidden audience whose exact size could only be estimated in the hundreds of thousands, or even millions. And it was a new audience as well, filled with a new generation that had never heard jazz and, in fact, had never heard of the Dixieland Band except in the lyrics of a recent Johnny Mercer song hit.

After a comic dialogue involving Wynn, announcer Graham McNamee, and Nick LaRocca, the band broke loose on a chorus of "Margie," and to some who remembered, the boys sounded better than they did on the original recording of the tune more than fifteen years earlier. Then, with Ed Wynn at the piano, they played one of the comedian's own compositions—"Swingin' In a Hammock for Two." (Wynn apologized that the number was originally called "Swingin' in a Hammock for One," but he was beside himself that year.)

But the thing that rocked the network that hot July night was the band's closing selection—the original number that had stunned the first audience at Reisenweber's in 1917, had horrified the reserved British in 1919, and had gone on to become the national anthem of the jazz musician—"Tiger Rag." The official fan mail count a few days later revealed that the Original Dixieland Jazz Band had done it again—they had pulled more listeners for the Ed Wynn show than the guest stars on fourteen previous programs. For ten minutes' work they were paid $750—as much as their whole week's salary at Reisenweber's twenty years earlier!

LaRocca was deluged with offers from all parts of the country, and now it seemed that the Original Dixieland Jazz Band was on the threshold of another great career—a whirlwind comeback that was at once spectacular, historic, and tragic.

18

LaRocca and the Nine Young Men

Successful as the original five-man combination had so far been, there was no doubt in the minds of the experts that the whole future of the Original Dixieland Jazz Band depended on its capacity for adjusting to the demands of the Swing Age. The novelty of hearing "Tiger Rag" played as only its composers could play it, of see- a "white-haired grandpop" like Larry Shields swing the pants off contemporary clarinetists, of seeing the first jazz band in action was unquestionably a tremendous attraction with great exploitable value. But although the small combo was an unqualified hit on guest appearances where its routine was limited to a special number plus a couple of encores, how well would the five men fare on a half-hour radio show or a four-hour dance program? Would dancers still dance for a whole evening to the rhythms of two decades ago?

It was clear to LaRocca that survival in the Swing Age necessitated conforming to the standard orchestration then in vogue and augmenting the band by at least another eight men. The original five members could then be starred as soloists at various times throughout the "big band" arrangements, or occasionally featured in their original style as a specialty act, in much the same manner as Goodman used his trio and quartet.

At the next rehearsal LaRocca outlined his bold new plan of converting the Dixieland Band into a fourteen-man swing orchestra. The project was an expensive one, involving the hiring of nine more musicians at union scale and obtaining the services of a top arranger. As Nick had already sunk nearly three thousand dollars into the reorganization of the original five, he now sought financial assistance among his compatriots. But although Robinson and Sbarbaro were quick to see the possibilities of another big-name band, Shields, as always, was flat broke and Edwards turned thumbs down on what he considered a risky speculation. In fact, he flatly refused to have anything to do with such a wild scheme.

So the fourteen-man outfit was formed without Edwards and began rehearsals during August in preparation for autumn bookings. The orchestration was identical with that of the Goodman and Dorsey bands: three saxophones (plus Shields) made up the front row, four trumpets (including LaRocca) and two trombones furnished the brass, and the rhythm section consisted of a piano, guitar, bass viol, and drums. Although Robinson and Sbarbaro played in the ensemble, LaRocca directed the orchestra for the most part with his trumpet tucked under his right arm, contributing only an occasional solo or hot lick as permitted by the written score. (This accounts for the total of four trumpets, as three were required for continuous service.)

Although he could not read music, Shields found many opportunities to embellish the ensembles with clarinet runs à la Goodman, and his presence in the combination gave it the "swing" flavor so richly desired.

It was just as well, however, that Edwards didn't go along, as his powerful tailgate would have been sorely out of place in the mechanical arrangements; the Dorsey-type high-register trickery so popular among swing trombonists of the new age undoubtedly would have confounded him. Although the "hot" clarinet style had changed relatively little during the last decade, perhaps nowhere was the style evolution more apparent than among the slip-horn artists, who had now come to rely more on the lip than upon the wrist.

The new Original Dixieland Jazz Band made itself publicly heard for the first time on September 2, 1936, in a recording session for Victor. The numbers chosen were "Tiger Rag" and "Bluin' the Blues," and here LaRocca had his chance to prove how little dance music had really changed over the years, and how modern and "swingy" his old jazz licks sounded when orchestrated for fourteen pieces. Actually this was LaRocca's sole purpose in staging the comeback, to claim his rightful due for the distinctive syncopated phrases that had come to be recognized as a product of the Swing Age.

The first recording session was followed by a second on September 9, when six more sides were grooved (see Table 8.) To ensure as wide an audience as possible among the teenage swing fans who now dominated the market with absolute and unremitting rule, the Victor Company saw to it that at least one of the Dixieland's numbers was issued back-to-back with the Benny Goodman band. "Clarinet Marmalade" was paired with Goodman's "St. Louis Blues." Hence the new Dixieland Band and Benny Goodman's orchestra walked arm in arm down the streamlined boulevards of the swing world, the Dixielanders capitalizing on the Goodman success while the Goodman crowd saluted the grand old quintet that had blazed the original trail.

Historically, however, the recordings made by the fourteen-piece Dixieland Band were not so important as the three subsequent platters grooved by the original five-man combination. These six sides, labeled The Original Dixieland Five, outsold the big band records and went a long way toward revealing to a new generation the musical magic that had lain dormant for two decades within the hearts of these great musicians.

These six masterpieces of the Original Dixieland Five were created at the final recording session of the big band. Although the Victor experts came out strongly in favor of adding a bass viol to the reduced combination, none insisted more strongly than LaRocca and Sbarbaro that the group record in its original configuration. How close the small band came to having a bass included is hinted in Sbarbaro's report that the bass player had already moved his instrument into the special studio while the heated discussion was taking place. It is indeed fortunate for posterity that the bass man was ultimately expelled from the studio, for the six sides as finally recorded demonstrate Sbarbaro's drumwork in far better fashion than would have been possible had he been forced to compete with another rhythm instrument.

The six numbers chosen for the new series were the same ones originally recorded in 1917 and 1918, affording an interesting comparison over the twenty-year period. Here again, for example, are the "Original Dixieland One-Step" and "Livery Stable (now Barnyard) Blues" back-to-back on a Victor record, just as

TABLE 8

Recordings of the Original Dixieland Jazz Band

(1936 Series)

By "The Original Dixieland Jazz Band" (Fourteen-Man Combination)

RECORDING DATE	COM-PANY	NUM-BER	TITLE	RELEASE DATE
September 2 1936	Victor	——	Original Dixieland One-Step	Not Released
September 2 1936	Victor	——	Satanic Blues	Not Released
September 2 1936	Victor	25403	Tiger Rag	Sept. 17, 1936
			Bluin' the Blues	
September 2 1936	Victor	25411	Clarinet Marmalade	*ca.* Sept. 25, 1936
			St. Louis Blues (Benny Goodman)	
September 25 1936	Victor	25420	Did You Mean It?	*ca.* Oct. 1936
			Who Loves You?	
September 25 1936	Victor	25460	Ostrich Walk	*ca.* Oct. 1936
			Toddlin' Blues	

September 25 1936	Victor	25668	Fidgety Feet Veini, Veini (Ronnie Monro)	*ca.* Oct. 1936
September 25 1936	Victor	——	Your Ideas Are My Ideas	Not Released
September 25 1936	Victor	26039	Old Joe Blade* Any Old Time At All (Lionel Hampton)	1936

By "The Original Dixieland Five" (Five-Man Combination)

October 9 1936	Victor	25502	Original Dixieland One-Step Barnyard Blues	1936
October 9 1936	Victor	25524	Tiger Rag Skeleton Jangle	1936
October 9 1936	Victor	25525	Clarinet Marmalade Bluin' the Blues	1936

*vocalist, J. Russel Robinson
Personnel: LaRocca (cornet), Edwards (trombone), Shields (clarinet), Robinson (piano), Sbarbaro (drums)

they were in the early days of mechanical recording immediately preceding the first World War. But although the changes, both technological and artistic, are vast, the ineffable and mysterious aesthetic element that always set the Original Dixieland Jazz Band apart from any other jazz band in history is still present in all its original vitality.

The greatest change, of course, is in the fidelity of reproduction. The mechanical brute-force techniques that rendered the first jazz record an acoustical nightmare have been replaced by the wonders of the blossoming electronics age—the microphone, vacuum-tube amplifier, and electronically-actuated cutting head. Instead of the funnel-shaped pickup horn each of the five musicians now plays into a separate microphone, the outputs balanced and mixed in the control room to assure equal volume for all instruments. The result is phonographic reproduction by which, for the first time, all members of the Dixieland Band can be heard at the same time— a far cry from the early mechanical recordings thoroughly dominated by Shields and Edwards. The melody —the bell-like tone of LaRocca's steadily driving horn —is now heard from the downbeat to the diminishing echo of the famous dixieland tag. While admittedly somewhat below the standards of present-day "hi-fi," the 1936 recordings nevertheless brought the sounds of the first jazz band into the modern era.

From the standpoint of the musicians themselves, the most outstanding advantage of electronic recording was their juxtaposition in the studio. In 1917 they had been forced to play while located several feet apart— the cornetist a full twenty feet from the pickup horn, the trombonist about fifteen, the pianist directly under it, and the clarinetist playing into his own individual side-horn. Now, placed in normal playing positions, they were able to hear one another and were therefore inclined to rely less on memory and take more liberty in their playing.

Artistically, the recorded performances were somewhat influenced by the Swing Age. The noodling clarinet solo, earmark of the contemporary swing band arrangement, was frequently worked in as the great Larry Shields took rides on whole choruses of "Original Dixieland One-Step," "Tiger Rag," and "Clarinet Marmalade." In fact, LaRocca is quick to explain that Shields was starred as the Dixieland Band's answer to Benny Goodman. The "Original Dixieland One-Step" solo, in particular, makes that answer plain and emphatic. Here Shields employs a wide variety of his famous "licks" (although by no means all) in a single chorus, as drummer Sbarbaro follows every move as flawlessly as if he were actually reading the clarinetist's mind. Snare drum rolls coincide with Larry's numerous runs and climax his final, high-pitched scream with a sureness that suggests a sort of sixth sense among these musicians. Certainly nowhere in any jazz performance has greater mutual understanding been exhibited by soloist and accompanist.

In "Tiger Rag" the clarinet breaks become more fluid and polyphonic than in the original version as Shields endeavors to keep pace with the tempo of the times—more "noodle" and less glissando. And in the famous "hone-ya-da" chorus, the clarinet part is brought forward as a solo instead of remaining in the background as an obbligato to the trumpet-trombone "riffs," as in the older records. Accordingly these trumpet-trombone figures are attenuated for use as a background to the solo. Thus, by comparing both Victor records of "Tiger Rag," we see the gradual evolution of a swing solo. It was this clarinet part of Shields, originally only intended as a background improvisation, that became more and more imitated throughout the 'twenties and finally worked its way into written arrangements as a standard routine.

Perhaps the most obvious change over the years is the slowdown in tempo. The blues numbers as well as the one-steps are rendered at a moderated pace, again more in keeping with the period. The effect is that of a more "stately" kind of dixieland band, as contrasted with the more uninhibited playing of five wild and reckless youths in 1917. The personnel is, of course, identical on both series, with the exception of J. Russel Robinson, who had replaced the deceased Henry Ragas on piano.

As to whether these men are playing better in their thoughtful middle age than in their unbridled youth may never be answered by comparing the two Victor series, as each has peculiar points in its favor. The consistency, spirit, and brilliant conception that characterized the early records must be weighed against the variety in arrangement and improved musical technique revealed in the new.

The brilliance of Edwards' tone seems to have become even more magnificent in the passage of years, and his rapid tonguing in the low register remains nothing short of sensational. Years of working out at the YMCA may have conditioned him to execute the exhausting bass-range solo in "Skeleton Jangle" without so much as a single gasp, although in the ensemble chorus his overzealousness has undoubtedly caused him to substitute a less impressive counterpoint for his old standard part—an unfortunate change. In the blues numbers, however, his harmonies are still extremely sensitive and beautifully conceived.

The trumpet of Nick LaRocca, so long obscured by antiquated recording methods, now stands out in bold relief in a perfectly balanced ensemble. We hear for the first time those rapid, fluttering lip-slurs in the first chorus of the "Original Dixieland One-Step"; those half-laughing, half-sobbing cries in "Barnyard Blues"; and the almost hysterical tone that drives "Tiger Rag" to its wild and clamorous end. The swift, beautifully-executed "incidental" runs in the verse of "Clarinet Marmalade" seem to roll off as if under their own power; while the melody of "Bluin' the Blues" is voiced with just about every consonant in the alphabet—trumpet tones that seem to begin with "y" and "l" as well as with the traditional "t." But whatever the emotion, whatever the effect, the LaRocca horn never once sacrifices its outstanding characteristic—that steady, relentless push that keeps the ensemble moving without a moment's hesitation from start to finish.

As always, it was the ensemble work that put the Original Dixieland Five in a class undeniably by itself. These musicians were designed for one another, depended on one another, and functioned only in their original combination. With all its parts well-oiled and properly adjusted, the Dixieland Five was a machine

that produced jazz with a precision and perfection un-
matched in history.

Nobody appreciated this more than Benny Goodman. During the recording session he sat entranced in
the control room, a wide grin on his youthful face. At
the conclusion of the last number he walked over to
LaRocca and remarked, "Nick, you boys have still got
something nobody else has got."

The Original Dixieland Five subsequently appeared as guests on Benny Goodman's weekly radio
broadcast. The applause following their selection lasted
nearly two minutes, after which Goodman remarked,
jokingly, "That's the last time I'll ever have them on
this program!"

Other radio performances followed in rapid succession—The Magic Key on September 20, Ben Bernie's
Blue Ribbon program, the Ken Murray Show, and on
October 31 the Saturday Night Swing Club. All of these
guest spots LaRocca had secured on his own, acting as
his own booking agent and hustling from one advertising
agency to another to maintain steady employment for
the band.

As the bigger jobs thinned out, however, Larry
Shields again grew moody. LaRocca had paid Shields'
way from New Orleans in July, including hotel bills
and living expenses until the band got working, and now
an argument ensued between the two of them concerning
financial readjustments. Larry considered the whole
thing part of LaRocca's business investment, and since
Nick was getting twice as much of the profit as anyone
else, didn't feel obligated to pay anything back. It was
a philosophical question, with each man sticking ob-

stinately to his own view. After a heated discussion Larry packed up and headed for home, and Nick called upon Artie Seaberg to take his place at Proctor's Theater, Schenectady, New York, on December 4.

In the meantime, Eli Oberstein of Victor had been instrumental in lining up a long-term vaudeville tour with Ken Murray, and Nick wanted the original band or nothing. Again he wrote to Shields, and on December 15 received this reply from New Orleans:

Dear Nick,
Received your letter stating the band has an offer to go on the road with Ken Murray. If this thing is positive and you can make arrangements to advance me railroad fare and enough money to live on until I make my first week salary, I can leave here whenever you say. Of course, it is understood I will pay this money back some each week as I work. Regards to yourself and Band.

Larry

Thus the Original Dixieland Jazz Band, perhaps the most temperamental group of musicians ever to exist, was back together and ready for the big events of 1937.

19

The Last Days of the Dixieland Five

The year 1937 was a tumultuous one for the Original Dixieland Jazz Band: a year of triumph, glory, conflict, and disaster. Their vaudeville tour of the country's leading theaters with the Ken Murray troupe was tremendously successful, beginning at the RKO Boston Theater on January 6 and carrying through the eastern, midwestern, and southern states during the course of two seasons. The sale of their Victor records also zoomed during this period, while the biggest publicity boost of all came through the *March of Time* film (Issue No. 7, Volume III, February 17, 1937).

In this movie, the band acted out its own history. The original recording equipment was dragged out of the Victor warehouse in Camden, New Jersey, reassembled, and the technician that originally recorded the band with this paraphernalia in 1917 appeared in the sequence where the Dixieland Band re-enacted the cutting of the first jazz phonograph record. (To preserve the illusion of youth, white-haired Larry Shields was kept out of camera range and the other musicians photographed from the rear.)

The film went on to show the deterioration of jazz music through the noisy imitators that followed, then brought the story up to date with a slightly embroidered version of the 1936 reorganization. LaRocca was shown combing the metropolis for his former bandsmen, picking them up one by one as he went from one building to another. Shields was located in a bookstore, loaded down with an armful of Bibles; Edwards appeared at the door of a gym locker room, wearing a YMCA sweat-shirt; Sbarbaro, in a suit of greasy overalls, crawled out from under a car; and the pipe-smoking Robinson broke off a radio rehearsal at NBC to join the march.

The story thus far was only half correct. Shields had been found in a Bible house, but not in New York City; and Sbarbaro had never seen the inside of a tire shop. The director of the film, Louis de Rochemont, had originally planned to show Sbarbaro washing dishes in a restaurant, but Tony protested so vehemently that an alternative scene had to be devised. Tony didn't like the tire shop idea, either, but it was better than washing dishes.

Following scenes portrayed the first rehearsal of the reorganized band, with LaRocca stopping the music to tell the boys to throw away their mutes, and concluded with an exciting night club sequence with close-up shots of Shields swinging out with his famous "Tiger Rag" solo. All in all, even if the story had been stretched a little here and there, it was a thrilling twenty minutes of dixieland.

One of the writers instrumental in the engineering of this unusual documentary work was J. S. Moynahan, also a jazz clarinetist, who was a long-time worshipper of the Dixieland Band. The experience was a disillusioning one for Moynahan, who objected to the distortion of historical fact insisted upon by de Rochemont in the interests of story unity. A particular sore spot was the portrayal of J. Russel Robinson as the band's original pianist, to the complete exclusion of the late Henry Ragas.

Moynahan followed up this project immediately with a feature article in the February 13, 1937, issue of *The Saturday Evening Post* entitled "From Ragtime to Swing," which was essentially a eulogy of the Original Dixieland Jazz Band. Some of his allegations raised a few eyebrows among the Goodman fans but carried an inescapable ring of truth:

> . . . there was only one Original Dixieland Jazz Band. And despite the floods of mystic adulation—not to say adumbration—it apparently takes more than a layman swing fan to explain, or even understand why. For that matter, I'm sure most musicians don't know. If they did, we wouldn't have had the plague of corny, McGee, ting-a-ling, strictly union recordings that spell "jazz" to the average customer. The most unspeakable butcherings of popular numbers in the name of jazz have, after all, been committed by musicians.
>
> What's the difference between swing and jazz? What is swing?
>
> A number of writers who, apparently, are not even musicians have been breaking into print lately with theses so esoteric that they become, at the high spots, practically unintelligible. It's not so hard as all that. In fact, it's simple.
>
> The difference between swing and jazz is, reduced to common honesty, nothing. And swing, despite the tons of recondite balderdash that have been printed about it, remains substantially what it was when grandpappy was a boy watching the band marching away to the Civil War, and remarked: "That music's got a swing to it."
>
> Swing is rhythm, that's all.
>
> The trouble with swing dates from the decline of the

Dixieland Jazz Band. This trouble, too, can be stripped of all nebulous theorizings and the salient element stated in a single word—syncopation. Swing, so-called, today, as a rule, hasn't enough syncopation. Father called it ragtime.

Those who are neither musicians nor old enough to know what ragtime is, may not understand what is meant by syncopation, or why it is so important in producing the superlative type of jazz . . .

The *March of Time* film, the *Saturday Evening Post* article, phonograph records, and lavish newspaper publicity all combined to ensure the success of the Dixieland Band's cross-country tour. With Harry Barth,[1] an old-time jazz man from Natchez, Mississippi, added on bass fiddle, the group scored new successes all the way from Boston to Chicago. In *Downbeat* for February, 1937, George Frazier, who had not been unduly impressed by the Victor records, rendered an eyewitness account of the Original Dixieland Jazz Band in action:

> The Original Dixieland Jazz Band in the flesh . . . is a decidedly more exciting proposition. First of all . . . it is one of the ensemble gems of jazz. Whatever else may be said of it, its cohesion and homogeneity are things quite rare and quite wonderful. Listening to LaRocca, Robinson, Edwards, Shields and Sbarbaro, you are aware that these guys work together with extraordinary sensitiveness. Their feeling for one another's playing and their ability to project that feeling into perfection are qualities frequently productive of unforgettable jazz. Shields . . . is very, very great. Amazingly enough, the years have failed to take their toll on his playing, so that today he sounds, if anything, better than he did in the past. Incidentally, it was only at the very last moment that he finally decided to

---

[1] Barth also played with the Original Dixieland Jazz Band for a few weeks during the early part of 1922, at the Balconades, doubling on string bass and tuba. The addition of a sixth member at that time was only an experiment.

join up for the current engagement, so that one shudders at the prospect of what would have been lacking had he remained at home in New Orleans.

In Chicago, at the end of the first Ken Murray tour, the band broke the circuit and headed south on their own. After a week at the Orpheum Theater in Memphis, where the words DIXIELAND JAZZ BAND were displayed on the marquee in letters three feet high, they continued south and arrived for a heroes' reception in their native New Orleans—the very city that had disowned them in 1918! In honor of their triumphant return, the management of the St. Charles Theater sponsored a banquet at Broussard's Restaurant. Among the guest speakers were George W. Healy, Jr., managing editor, and other members of the staff of the *Times-Picayune*, the newspaper that had, twenty years earlier, excoriated the originators of jazz with its "Jass and Jassism" editorial (see Chapter 6). Speaking at the dinner, A. Miles Pratt, president of the St. Charles Theater Corporation, expressed the hope that New Orleanians would "forget their past efforts to disclaim the parenthood of jazz and welcome the band back home in proper spirit." This hope was confirmed that week when the Original Dixieland Jazz Band, playing to overflowing crowds at the St. Charles, created such a traffic jam that cars had to be rerouted around the block. Later they were guests of the New Orleans Association of Commerce at another luncheon at the Roosevelt Hotel.

It would seem odd that with everything going their way, the members should choose to begin a quarrel that would jeopardize their very future. But if money is "the

root of all evil" the ground was fertile, and the arrangement whereby LaRocca divided the profits into six parts
and kept two parts for himself (with Barth on straight
salary) was sure to engender savage resentment in the
end. Although this arrangement was agreed upon by all
at the outset of the reorganization and obviously seemed
justified in consideration of LaRocca's sizeable initial
investment, at least one member now paused to reflect a
situation in which he played as much music as Nick but
only earned half as much money. In this act of rational-
ization, it was all too easy to forget LaRocca's financial
risk, as well as the losses incurred in closing up his New
Orleans business.

It may be said that LaRocca suffered a disadvan-
tage common to all who rise from the "ranks." The
Original Dixieland Jazz Band had been a co-operative
organization in the early days, with the offices of musical
leader (LaRocca) and business manager (Edwards)
considered merely additional chores. Even after the cornetist took over the responsibilities of business manager,
the outfit functioned as a financially co-operative unit.
Signatures appeared in no established order on any of
the contracts prior to 1920, apparently the member nearest the pen being the one to sign on the top line. Likewise, all royalties, profits, and even composers' credits
were shared alike. It was not until after the band's return
from England that LaRocca, as stated manager, signed
for the entire organization. Therefore, one may surmise
that LaRocca's rise to the position of sole and exclusive
authority was a gradual evolution, resulting from the
sheer necessity of leadership under increasingly trying
conditions in the music business.

If this be true, then Edwards, Sbarbaro, and Shields were two decades late in their demand for a "vote of confidence." The band had already been through its most perilous storms under LaRocca's direction, and there would seem little justification for the anarchy they proposed as its substitute.

Perhaps the situation was irritated by a growing mistrust of the LaRocca-Robinson relationship. On October 8, 1936, the two composers had filed for a license to conduct a business at 308 West 58th Street in New York City, a publishing concern to be known as Original Dixieland Music. Under this set-up, LaRocca and Robinson planned a series of popular compositions to be plugged by the band and published by their own company. The first of these was "Old Joe Blade" ("Sharp as a Razor"), a song based on the "Casey Jones" vehicle and very similar in structure. "Old Joe Blade" was introduced at college dances late in the fall of 1936 and achieved a small measure of success among those who were inclined to sing along with the band. LaRocca believes that the other members were suspicious of the alliance between himself and Robinson and wrongly assumed that they were plotting to appropriate more than their own share of the profits.

Now, in New Orleans, these grievances came to the surface. Shields had been morose and uncommunicative for some time. "Between sets," recalls LaRocca, "he would sit there with the clarinet on his lap, just staring into space." Larry, fighting his depression, had begun to drink perhaps more heavily than was his custom, and although good-natured and easygoing, he was easily influenced. In this frame of mind he was particularly susceptible to the ideas of others. As the story goes,[2] Edwards and Sbarbaro treated Shields to a few rounds of drinks one night and won him over to their side, thus beginning a conspiracy that eventually led to the breakup of the band.

Prompted by his compatriots, Shields requested a business meeting for the purpose of reviewing the existing financial arrangements and seeking "a better break" for himself and the others. Although there is no record of what transpired at the meeting, it is safe to assume that Shields' position of "starred" soloist in the new band was used as a strong selling point. Nevertheless, the meeting was doomed to end in a deadlock. When LaRocca asked who was to reimburse him for his initial investment, not a voice was heard.

And so, with the matter obviously settled for the nonce, the Dixieland Band resumed its tour of the northern states in March, playing movie theaters from Milwaukee to New York. But the bitterness of the three men was not diminished, and it is natural to suspect that some plotting was still afoot. LaRocca asserts that in Cincinnati the disgruntled members deserted him at the crucial moment of curtain time, leaving him embarrassingly on stage without a band. Although the others deny this or find themselves unable to remember, Sbarbaro reveals that he knew the band was going to break up and was therefore keeping up his outside contacts. Robinson, with a note of disgust, says that they were

[2] The following account of the band's breakup is based on personal interviews with LaRocca, Edwards, and Sbarbaro during 1957 and 1958.

always quibbling about something and that he did his best to stay out of it.

On April 21 the band began a two-week engagement at the Silver Grill in Buffalo. Outwardly there were no signs of the dissention that tore at the organization from within. While LaRocca and Edwards discussed the placement of the band on the small platform, Tony Sbarbaro sat at a table, laughing and reminiscing with a couple of old-timers who in 1917 had driven all the way from Buffalo to New York, fighting four hundred miles of muddy ruts to reach Reisenweber's to hear the famous Dixieland "Jass" Band in person. Now they asked him about the nodding teddy bear that used to sit atop his bass drum in the old days, and Tony retaliated with questions about this strange town, where half the population seemed jammed into a bar about the size of a New York cloakroom for the chance to hear the first real jazz ever to come their way.

Then LaRocca ripped off a few bars of "At the Jazz Band Ball" on his trumpet, and the bandsmen emerged from the crowd to begin their first number. The old-timers listened as if under a spell, silently comparing the Dixieland Band of 1937 with that of 1917. They marvelled again at the apparent ease with which this music was produced, and the softness that always surprised people who had never heard them in person. In reality the Dixieland Band was a mere whisper as compared with the blaring swing bands that typified the present trend in dance music. But all is relative. In 1917, audiences that had become accustomed to string ensembles were shocked by the intrusion of parade band instruments and were quick to describe the radical new music as "deafening." More startling, in fact, to its new listeners in the Swing Age was the extreme ease and powerful syncopation of the New Orleans band. The likes of this they had never heard before.

LaRocca, Edwards, and Shields blended together as well or better than ever. The old-timers at the ringside table thought LaRocca slightly more tense than usual, Edwards the more relaxed. Robinson played with a certain air of detachment, contributing his usual full foundation to the ensemble but maintaining a cold and stony expression even during the most thrilling and inspiring choruses. As for the audience, they were too excited to applaud. They bellowed their approval.

On the surface it may seem paradoxical that men incapable of a harmonious working relationship were responsible for some of the greatest ensemble jazz in history, since a high level of understanding would naturally appear essential. Yet even in an argument there is both co-operation and understanding. The most violent combatants are quick to answer each other's questions, ready to seize every opportunity for the expression of their emotions. The musicians of the Original Dixieland Jazz Band were artists of highly sensitive temperaments, individualists of stubborn bent. It was this extreme individualism, after all, that had given birth to the form of music called jazz in the first place. Each musician expressed his own ideas in the form of musical counterpoint. When LaRocca made his pre-emptory statement on cornet, Shields noodled his comment via the clarinet and Edwards snarled a forceful reply on trombone. Here there was no need for love and admiration, only a sensitive feeling for one another on a strictly musical plane.

After an appearance with Xavier Cugat's orchestra on the stage of New York's Paramount Theater in May, the Dixieland Band again took to the road, arriving in Fort Worth on June 26 for the opening of Billy Rose's widely heralded Frontier Fiesta. It was at the Fort Worth exposition that the rebellion within the famous jazz band reached its most dangerous proportions.

The band had been engaged for the Pioneer Palace Review, a musical pageant presented on a circular outdoor stage divided into several pie-shaped sets. Many acts were in progress simultaneously, as the roving audience passed from one set to another in its quest for entertainment. LaRocca maintains that Robinson accepted a job on one of the other sets in an act called The Great Composers, hiring a pianist to take his place temporarily in the Dixieland Band and thereby drawing two salaries. The other band members, suspecting a "deal" between LaRocca and Robinson, objected vehemently and accused the cornetist of collusion. Undoubtedly LaRocca was caught in the middle, as Billy Rose had personally arranged the switch and the band was solidly under contract for the season.

The crisis mounted when a sound truck, hired by LaRocca, cruised about the fairground advertising "Nick LaRocca's Original Dixieland Jazz Band." Edwards saw red. In his mind the Dixieland Band had always been a co-operative organization with no member billed above another, and now the trombonist gritted his teeth as LaRocca's name went booming across the countryside. That night, as the mutiny progressed into its

last stages, Edwards and Shields diabolically played off
to the side so that only the sound of LaRocca's cornet
was picked up by the microphone. If Nick is the
leader, argued Edwards, let him have the whole stage
to himself.

And so the war of nerves continued, the campaign
to wear down LaRocca through public embarrassment.
But as the prestige of the band suffered accordingly, it
would almost seem that the plan was more vindictive
than practical.

Meanwhile another phase was unfolding behind
the scenes. LaRocca claims that his men had gone to the
local office of the musicians' union and had convinced
them that he was not even a member of the Original
Dixieland Jazz Band! Incredible as it seems, the following letter from the president of Fort Worth's Local 72,
dated October 11, 1937, appears to lend credence to
the story:

Dear Nick:

I am enclosing your receipt for tax. Here's hoping that
you are well and healthy and have a chance to join the band. I
have always thought that you would make a big success if you
could only become a member of the original Dixie Land Jazz
Band.

Fraternally yours,
Woods C. Moore, President
Local 72, Am. Fed. of Musicians

As LaRocca explains it, he then visited Local 72
to determine the meaning of this enigmatic letter. Moore
revealed the plot that had been brewing and advised
LaRocca to bring charges of conspiracy against a leader.

With the interest of the band at heart, Nick declined the suggestion.

The Original Dixieland Jazz Band rejoined the Ken Murray troupe in November, and for awhile the situation seemed under control. Then, while the band was in New York, someone cracked LaRocca's trunk in the basement of the Wilson Hotel and made off with many valuable items and records, including the corporation seal of the Original Dixieland Jazz Band, Inc., and all written arrangements for the fourteen-piece orchestra. Tensions increased throughout December, as LaRocca overheard threatening conversations, either real or imaginary, in adjoining rooms of his hotel suite. On-stage co-operation again reached a low ebb.

The cornetist warned his men that his position in the band had at last become untenable and that unless the situation was immediately rectified he would disband his outfit and return to New Orleans. The threat was met with "razzberries" and horselaughs from the bandsmen, who thought they knew Nick well enough to be sure that he would never quit a money-making proposition.

It was Monday night, January 17, 1938. Wild crowds at Chicago's Palace Theater were still applauding, yelling, and whistling as the curtain rolled down on the Dixieland Band. Larry Shields, who had just brought down the house with his famous "Tiger Rag" solo, was fitting the warm segments of his clarinet into their oblong box, while Edwards blew the juice out of his trombone and Sbarbaro, on the top step of the bandstand, loosened the screws on his bass drum. Ken Murray stood smiling in the wings, but LaRocca's face was

grim. Reaching into the pockets of his jacket, he produced five pieces of paper and handed them to the
members of his band. The papers bore the following
words:

> Palace Theater
> Chicago, Ill.
> Jan. 17, 1938

A. Sbarbaro, L. Shields, E.B. Edwards, et al.
Palace Theater
Chicago, Ill.
Gentlemen:—

After the completion of the present engagements with
Ken Murray, I hereby give notice that the Original Dixieland
Jazz Band will be disbanded.

I am very sorry to have to come to this conclusion, but
owing to the internal friction, which makes it impossible to
carry on, I am mailing a copy of this notice to Local #802.

> Very truly yours,
> *D. Jas. LaRocca, Leader and Re-organizer*
> Original Dixieland Jazz Band

Upon threat of legal action from the agents, the
group was compelled to finish out its present commitments. But on February 1, as the last notes of its famous
music diminished to an absolute silence, the Original
Dixieland Jazz Band was dead.

The consequences were not immediately realized,
especially by the band members themselves. LaRocca
believed he had done his job, proved his point, stamped
his mark indelibly in musical history. The others thought
they could carry on without him. Both were wrong.

Just as in 1925, when LaRocca left the band it
immediately dropped to the level of a rather freakish
curiosity—a band with only novelty value. Under the
"co-operative" leadership of Shields, Edwards, and

Sbarbaro, the combination faltered in attempts at compromising dixieland jazz with commercial swing. Names never meant anything in the Dixieland Band as far as the average jazz fan was concerned. A cornetist who played half as well as LaRocca could have enabled the band to carry on—few would have missed that magical name—the name that had echoed across the Fort Worth fairgrounds in July. But no such cornetist was in existence, and unless the remaining members of the Dixieland Band were totally blinded by their own importance they must have known it as well as anyone else. This may have been why they immediately deserted their natural style for swing.

Yes, here was one substitution that could not be made. Although he himself had successfully substituted a new drummer in 1916, a trombonist in 1918, a pianist in 1919, and a clarinetist in 1922, LaRocca proved for the second time in a dozen years that his own substitution was impractical.

As for LaRocca's great mission, it was not so thoroughly accomplished as he had thought. He was correct in his belief that he had reached the people, and those who came to listen never forgot. But these cheering thousands were but an infinitesimal portion of the population. LaRocca had misjudged; the job had only begun.

Edwards, Shields, and Sbarbaro subsequently reorganized the Dixieland Band as a swing outfit, with Sharkey Bonano on trumpet and Frank Signorelli on piano. Sharkey by now had absorbed some of the Harry James flamboyance that permeated the swing world, and Signorelli had long since undergone a complete transformation. One could hardly believe that this so-phisticated concert-hall style emanated from the same pianist who in the early twenties pounded out those great ten-fingered foundations for the Dixieland Band and the Memphis Five.

In 1938, identified as "The Original Dixieland Jazz Band, with Shields, Edwards and Sbarbaro," this group recorded six sides for Bluebird. (See Table 9.) Popular tunes of the day were rendered in a bastard style that attempted to straddle the never-never land be-tween dixieland and swing, aided by a tremulous girl vocalist named Lola Bard. The records, however, are interesting as a study of Larry Shields' remarkable tech-nique. The Shields clarinet is heard continually, and the easy-going, freely-flowing obbligato he furnishes for the vocals is well worthy of careful attention.

There are indications that at this time Edwards was doing everything in his power to patch up his shattered friendship with LaRocca. In July of that year he wrote to Nick in hopes of reorganizing the old band for a European engagement that had been offered him. But Nick was far from approachable on any subject con-nected with music. He replied with bitter memories: ". . .what has transpired in Chicago and Cincinnati I have already informed Local 802, but on second thought I would not try and deprive any of you from trying to make a living, even as low as you fellows were to me. For your information the charge would be conspiracy on a leader. . . . I have given two best years of my life and home happiness to bring you fellows back to fame and fortune, and all I received from you was abuse and in-sults. Some gratitude for my efforts."

TABLE 9

Recordings of the Original Dixieland Jazz Band
(1938 Series)

COMPANY	NUMBER	TITLE
Bluebird	B-7442	oooOO-OH Boom!
		Please Be Kind
Bluebird	B-7444	Good-night, Sweet Dreams, Good-night
		In My Little Red Book
Bluebird	B-7454	Drop a Nickel in the Slot
		Jezebel

Personnel: Sharkey Bonano (trumpet), Eddie Edwards (trombone), Larry Shields (clarinet), Frank Signorelli (piano), Tony Sbarbaro (drums)

As late as 1940 the Shields-Edwards-Sbarbaro team was still active around New York. On January 16, with J. Russel Robinson on piano, they made a guest appearance on the CBS radio show We the People. Considering the assortment that ordinarily found expression on this program, however, there was little to be said for this latest achievement of the "Original Dixieland Jazz Band." LaRocca wisely refused the use of his name in this petty come-down, especially since the facts were so flagrantly distorted in the radio script.

Larry Shields returned to California to take a defense job in a shipyard, but in 1943 another Original Dixieland Jazz Band sprouted up on Broadway in Katherine Dunham's Tropical Revue, followed by a road tour with the same company. Edwards and Sbarbaro were the main features of the reorganization, with Bobby Hackett on trumpet and Brad Gowans on clarinet. A V-disc of "Tiger Rag" and "Sensation Rag," recorded for the armed forces, exists as a monument to this effort.

Of all the subsequent attempts, this particular combination comes closest of all to the spirit of the original five.

But now, once again, hot music and small combinations were on the way out. By the end of the war all big bands attempted conformance to the Glenn Miller pattern, Tommy Dorsey had given up the Clambake Seven, Bob Crosby had forgotten about his famous Bobcats, and real jazz retreated to the Bohemian caves of Chicago and Greenwich Village.

20

Syncopated Echoes

One of the questions most likely to be asked by future jazz historians is why, during 1949 and 1950, when dixieland was enjoying its greatest popularity in more than thirty years, did the Original Dixieland Jazz Band not return to the scene to re-establish its rightful claim? With every old-time musician who had ever sounded a note of jazz making a profitable comeback, certainly the band that had started it all would again become a sure-fire public attraction. The original members—with the exception of pianist Henry Ragas—were still very much alive and just as capable of staging a spectacular reappearance as they were in 1936.

That the Original Dixieland Band did not return in the favorable environment of the dixieland renaissance is surely one of the most tragic misfortunes of the period. These men could have placed their stamp on a whole new generation of jazz fans, and with actions rather than words. Even more important, they could have recorded their music with the technically improved facilities of the high-fidelity age, leaving for posterity a series of recordings faithfully covering the entire tonal range of their artistic product.

The reasons for this unhappy silence are as simple as they are hard to justify. The members of the Dixieland Band were highly temperamental musicians, easily given to the selfishness, jealousy, and egotism so often a part of the artistic makeup. While profit in dollars and cents was often the only thing that held them together, personal rivalry was more frequently the deciding factor in their collapse. More interested in individual glory than in team recognition, they blindly destroyed the vehicle that had made them what they were. Of the four surviving members, perhaps only LaRocca was aware of the supreme importance of their interrelationship. Even today, in spite of the animosity that continually mounts among them, he never fails to give each of his former musicians credit for being the greatest in his line, a gesture which unfortunately is rarely reciprocated.

J. Russel Robinson, who contributed such a notable part in the band's last efforts, firmly believes that the Original Dixieland Band had immortality within its grasp, an opportunity lost forever through the personal squabbles of its members. Such immortality will, of course, eventually be achieved, but only after the smoke of the present internecine warfare has cleared.

One of the deepest schisms in the band's personnel relationship was the matter of royalties accruing from group compositions, shrewdly copyrighted by agent Max Hart in his own name in 1917. The wound was opened anew in 1943, when the copyrights came up for renewal. By proving himself the band's leader and showing that royalty checks had always been made payable to the band at his 2022 Magazine Street address in New Orleans, LaRocca was able to straighten out the matter with the copyright office. Then, in consideration of an outright cash settlement, plus royalties to himself and the original members or their heirs, he transferred all rights to the publishers, Leo Feist and Company. The case is too involved for coverage in a single volume of jazz history, but the matter seems not to have been settled to everyone's satisfaction.

Ill feeling among the musicians was furthered by the ASCAP affair. Shields had come back to New Orleans a few years after the band's final breakup to apologize to LaRocca for the part he had played in the Fort Worth conspiracy, and to seek the bandleader's help in becoming a member of the American Society of Composers, Authors, and Publishers. LaRocca had joined the society in 1937 and was happy to secure a membership for his former collaborator, based on Larry's contributions to such Original Dixieland tunes as "Clarinet Marmalade," "Ostrich Walk," and "Look at 'em Doin' It." But when Edwards and Sbarbaro wrote to LaRocca for the same purpose, hoping to gain his consent to come into the organization as co-authors of "Tiger Rag," "Sensation Rag," and "Mournin' Blues," their pleas quite naturally met with a stony silence. Hence, friction among the band members reached a new peak.

In 1947 Milt Gabler of Commodore Records decided that conditions in the music world were right for another revival of the historic combination. LaRocca declined and Shields, in California, was too far removed; but a few local telephone calls found Edwards at his old haunt, the West Side YMCA on 63rd Street. Edwards had been sporadically active in music around town since the revival of hot jazz. Now he borrowed Tony Sbarbaro from the Memphis Five long enough to record a ten-inch LP entitled *Eddie Edwards and his Original Dixieland Jazz Band* (Commodore FL 20, 0003). Although the net result is extremely disappointing when compared with the Original Dixieland records of yore, these sessions produced the last recorded sounds of Edwards' great trombone and deserve careful notation for that, if for no other reason.

As to be expected, the most outstanding feature of these performances is Edwards himself, who, even as he approaches the age of sixty, demonstrates once again a tailgate power surpassed by few others on the contemporary scene. But the trumpet of Wild Bill Davison is more suited to solo jazz and in most of these old dixieland numbers it is just too wild for controlled ensemble work. Edwards, apparently confused by the explosive outbursts of Davison and crippled by the lack of a dependable lead like LaRocca's, is unable to knit his phrases together, unable to achieve the cohesion and unity of counterpoint that helped ensure the fame of the old Dixieland Band.

Brad Gowans offers a pleasing imitation of Shields' famous clarinet breaks in "Tiger Rag" and "Ostrich Walk," and his tone and technique are at times startlingly close to that of his idol; yet in the ensembles his clarinet does not cry out in the old Shields tradition. The combination of Tony Sbarbaro (drums), Teddy Roy (piano), Jack Lesberg (bass), and Eddie Condon (guitar) generates a good beat, but after the opening chorus, nearly every opus deteriorates into a typical Condon-type jam session—replete with all those lifeless piano choruses and uninspired "one-two-three" solo routines.

The Commodore idea of mixing solo jazz stars with a couple of genuine ensemble men was not the answer. This would never be the Original Dixieland Jazz Band as it had existed under LaRocca. But the chances of reorganizing the original five were slight. LaRocca was busily engaged in the contracting business in New Orleans, building houses and supporting a family of eight. No amount of pleading could get him back on the road. Nor was there any chance of patching up the differences that still existed between LaRocca and his former bandsmen.

Then, in Hollywood, even the remotest possibility of such a reorganization was extinguished for all time. On November 23, 1953, the great Larry Shields passed away, the victim of a severe heart attack. Across the country, in a Forest Hills, New York, apartment, Tony Sbarbaro sat stunned at his radio loudspeaker as the voice of Walter Winchell conveyed the news.

The world may never know whether it was the clarinet, or his having to give it up, that killed Larry Shields. Shortly after Larry returned to California after the demise of the Original Dixieland Jazz Band in 1938, a critical heart condition became apparent and his doctor advised him against ever playing the clarinet again. But this was like telling a fish to stay out of the water. For a man whose musical instrument was his very life, such advice was far more easily given than heeded. The white-haired clarinetist continued to play occasional jobs. But his face would become as red as a lobster whenever he took a hot chorus on his clarinet, and his near-frantic wife finally succeeded in convincing him that total retirement from music was his only salvation.

During World War II Larry went to work at the shipyards, working on blueprints. Mrs. Shields, who still lives in Hollywood, recalls how she and her husband would meet musicians while strolling on Hollywood Boulevard, and how they would plead with him to sit in for an evening. Each offer refused became a new self-wound for the retired clarinetist. On one occasion a special jazz concert was being arranged for the annual Shriners' affair, featuring all the top jazzmen of the West Coast, and Larry was asked to appear. They even wanted his name in lights. When Larry declined, they asked if he would just stop backstage and say "hello." But Larry never came around.

Now there remained but three surviving members of the "original" Original Dixieland Jazz Band to defend themselves against the careless misrepresentation of jazz history that was taking place all around them. On September 5, 1954, over the CBS television network, came an installment of the You Are There series, a documentary-type episode entitled "The Emergence of Jazz," which was purportedly an authentic account of the evolution of this music. Depicting, on this program, the members of the Original Dixieland Jazz Band, were five New York jazzmen, including Bobby Hackett (as Nick LaRocca) and Lou McGarrity (as Eddie Edwards). These musicians signified their presence by playing a few notes on their respective instruments as the narrator called the roll. As the story unfolded, a pianist representing Jelly Roll Morton claimed that he had written "Tiger Rag" and that the Original Dixieland Band were "thieves and miscreants" for appropriating a number that was rightfully his. This, of course, was merely an

elaboration on the myth that Morton had already per-
petrated on a series of records made for the Library of
Congress more than a decade earlier.

Edwards was jolted into action. The aging trombonist immediately engaged the services of a New York
legal firm to bring suit against the Columbia Broadcast-
ing System and Walter Kronkite for unauthorized use of
his name and those of his fellow band members. The
right of the program to impersonate historical personalities was fairly well established through precedent; but
Edwards' complaint that none of the living members of
the Dixieland Band had been contacted for permission,
even though their names were still listed on the roster of
Local 802, seemed valid. Bobby Hackett and Lou Mc-
Garrity, both friends of Edwards, felt that a gross in-
justice had been inflicted upon the Dixieland Band and
agreed to testify in their behalf. Likewise, Tony Sbarbaro was quick to join battle. The $100,000 suit was
eventually settled out of court.

Meanwhile, far from the limelight of the entertainment world, a familiar, dissatisfied rumbling was beginning to be heard. As with every "hot" jazz revival, the
voice of the grand old man of dixieland could not remain silent. Nick LaRocca, his cornet tucked away in its
battered case for ten peaceful years, re-entered the scene
in a new role: that of anti-critic. Having effectively
stifled the skeptics in 1936 with a successful reorganization of his old band, he now returned as an extremely
vitriolic critic. Armed with a typewriter, a garage full
of weathered documents, and a newly purchased photo-
stating machine, the ex-bandleader launched his war
with courage and conviction. Newspaper columnists,

authors, jazz club officials, radio networks—none who confused the provable facts of jazz history were safe. The letters, although shocking in their unrestrained vernacular, in many cases hit their mark. Whatever else may be said of the LaRocca crusade, it must be admitted that he gave no quarter, made not the smallest compromise with truth, even when it might have advanced his cause.

One of Nick's letters to the New Orleans *Times-Picayune*, complaining about a feature article on jazz history that had neglectfully overlooked the Original Dixieland Band, brought a newspaper reporter and photographer to the LaRocca residence. LaRocca's side of the story followed in the Sunday issue of August 3, 1958, together with a three-column photograph of the irate musician. The article started an avalanche of mail from old friends and acquaintances, some of them unheard from in thirty or forty years. Among these was Phil Napoleon, whose letter of August 21 read in part:

Dear Nick,

Some folks who live on the street next to us here in North Miami . . . have sent me the story out of the Times-Picayune. Well, needless to say, I was ever so happy to get this, but more than just reading it over and over must admit that I got such a satisfaction from its contents that I just had to get this off to you. Now, Nick, only I and a few others who are still alive can and will swear and confirm and help prove that every word of what you gave that reporter is the truth, so help me God. Again might I add that we of the Original Memphis Five . . . Frank Signorelli, piano; Miff Mole, trombone; Bill Lambert, drums; . . . Angelo Schiro, clarinet . . . are so mad over this entire thing and want you to know that. Now may I make this clear, that in those lean days we the Memphis Five had only the Dixieland Band to copy from and try to play the many things you all gave the world, for without your band how could we have been able to make the little success we did? There were no records of any worth to go by in those times— that is, in the way of true Dixieland music as we know it to-day. . . . So I am here to go to any extent to help you in this cause, to prove to the world that your band was the first, and all the rest of us climbed on to something that still is rightfully yours . . .

It was your band who took this music (that they often try to place in a bad house or spot of those days), who played in the finest night clubs the world over—and, by the way, didn't drag it down with drinking whiskey on the stand and those funny kinds of things they smoke—no, none of that, but I do know that the band on so many occasions was wearing white tie and tails in order to uphold their feeling for the good music they conveyed, and in the best and smartest of places . . .

It was the Dixieland Band who ventured up many times, heartsick in their attempts to sell something that was new and different . . .

Nick, I could go on and on, but please trust me when I say that I feel deeply for what you must sustain and the hurt things like this can bring to someone such as you who has given so much to musicians all over the world . . .

Phil

Phil Napoleon had left the music world in 1956 to spend his well-earned retirement on the shores of Miami Beach. Billy Maxted, Phil's former pianist-arranger, and his Manhattan Jazz Band took over at Nick's Restaurant in the Village for two very successful years, carrying on in the old Memphis Five tradition.

Napoleon is sincerely working for a reconciliation of the surviving members of the Original Dixieland Jazz Band, and his admonition of "Boys, this is no way to go out!" best expresses the thoughts of others who have tried. But the task he has cut out for himself is one of monumental proportions. Stubbornly opposed to any curtailment of the LaRocca-Edwards feud is Edwards himself, who has become extremely vocal on the subject in recent years.

Nick LaRocca lives today in a two-story house of his own construction at 2218 Constance Street in New Orleans, in the same neighborhood that in 1897 heard the first baleful sounds of his cornet. His was an inventive, creative, industrious nature that would have ensured success in any chosen line. Several houses in the Magazine Street neighborhood stand as monuments to his workmanship, as do miscellaneous restaurants, garages, and other buildings he has constructed singlehandedly.

Although now fully retired from business, the idea of idleness is as unthinkable to Nick LaRocca today as it has been all his life. The creative energy which gave nineteen jazz classics to the world now channels itself back into music. Under doctor's orders, the cornet will never be heard again; but the mind that cannot tell one written musical note from another continues to develop new themes. Jazz numbers such as "Irish Channel Drag," "Down in Old New Orleans," "Let's Jam It," "Is it True," "Now Everybody Step," and "You Name It" came into being during 1958, as did several ballads and waltzes. A clever bit of verse by Ed T. Jones was set to music by LaRocca that same year and became known as "Swamp Water Ballad," now in the process of publication.

The LaRocca method of musical composition remains the same as always—melodic improvisation upon the chord progression of one of his previous works. He composes at the piano, pounding out the chords with a heavy touch and humming or whistling the melody. The chord symbols, which he has come to recognize through long experience, are scribbled on the backs of old envelopes or any scrap of paper at hand; but the melody is carried in his head until son Jimmy comes around to capture it on staff paper. The melody comes out slightly different each time it is hummed or whistled, so the number often passes through many stages of evolution between its conception and the moment of transcription.

The rhythmic ideas, phrases, and tonal effects developed by Nick LaRocca nearly a half century ago are likely to remain fundamental to jazz for many years to come, regardless of its modern directions. But his impact upon American music may not be fully appreciated until jazz is more fully understood by more people. It is a music that took its orchestration and basic time from the parade bands of New Orleans, its harmonic structure and patterns of composition from the European tradition, and its melodic improvisation from the particular brand of ragtime that flourished in every seaport town from the Crescent City to the Barbary Coast. But something new was added: a most revolutionary style of hitting before or after the beat—a distinctive *syncopation* that is the essence of jazz. The whole jazz ensemble revolved around this characteristic approach to the melody —the trombone answering the cornet in a voice strongly reminiscent of military bands, and the clarinet developing an entirely new "noodling" style to fill in the gaps left for this purpose in the melodic phrasing.

As jazz goes into the fifth decade of its history, the fashions in that contagious form of syncopation continue to change. On through the pages of history march its heroes—Whiteman, Goodman, Armstrong, Gillespie— each hailed in his own day. But lingering like ghosts in the background—first ridiculed, then widely acclaimed, later attacked, and finally all but forgotten—are Nick LaRocca and his musicians of another age, the five pioneers who brought into existence the most phenomenal revolution in the annals of American music.

TABLE OF PERSONNEL
Original Dixieland Jazz Band

DATE		PLACE	CORNET	TROM-BONE	CLARI-NET	PIANO	DRUMS
Oct.	1915	Canal & Royal, N.O.	LaRocca	Mello	Nunez	—	Laine
Oct.	1915	Haymarket Café, N.O.	"	"	"	Ragas	Stein
March 1	1916	Schiller's, Chicago	"	Edwards	"	"	"
June 2	1916	Del'Abe Café, Chicago	"	"	"	"	Carter
June 16	1916	Del'Abe Café, Chicago	"	"	"	"	Sbarbaro
Nov. 3	1916	Casino Gardens, Chicago	"	"	Shields	"	"
Sept. 7	1918	Reisenweber's, N.Y.C.	"	Christian	"	"	"
Dec.	1918	Reisenweber's, N.Y.C.	"	"	"	Lancefield	"
Jan.	1919	Reisenweber's, N.Y.C.	"	"	"	Robinson	"
Oct. 11	1919	Rector's, London	"	"	"	Jones	"

Date	Year	Location					
Sept. 1	1920	Folies Bergère, N.Y.C.	"	Edwards	"	Robinson	"
April 11	1921	Folies Bergère, N.Y.C.	"	"	"	Signorelli	"
Dec. 24	1921	Balconades, N.Y.C.	"	Costello	"	"	"
Jan.	1922	Balconades, N.Y.C.	"	Lytel	"	"	"
April 10	1922	Flatbush Theater, Brooklyn	"	"	Seaberg	Vanicelli	"
ca. June	1926	Cinderella, N.Y.C.	Levine	Kaplan	"	Chase	"
July 28	1936	NBC Studios, N.Y.C.	LaRocca	Edwards	Shields	Robinson	"
ca. March	1938	New York City	Bonano	Edwards	Shields	Signorelli	"

NOTE: The first three combinations listed were **not** billed as the Original Dixieland Jazz Band and are included for reference only. All others carried this billing.

Benny Krueger (saxophone) was added in 1920 for recording purposes only. Harry Barth (string bass) was added in 1922 for a brief experiment, and again in 1937 for the road tour only.

Index